WEYERHAEUSER ENVIRONMENTAL CLASSICS

Paul S. Sutter, Editor

WEYERHAEUSER ENVIRONMENTAL CLASSICS
Paul S. Sutter, Editor

Weyerhaeuser Environmental Classics are reprinted editions of key works that explore human relationships with natural environments in all their variety and complexity. Drawn from many disciplines, they examine how natural systems affect human communities, how people affect the environments of which they are a part, and how different cultural conceptions of nature powerfully shape our sense of the world around us. These are books about the environment that continue to offer profound insights about the human place in nature.

Making Climate Change History: Primary Sources from Global Warming's Past, edited by Joshua P. Howe

Nuclear Reactions: Documenting American Encounters with Nuclear Energy, edited by James W. Feldman

The Wilderness Writings of Howard Zahniser, edited by Mark Harvey

The Environmental Moment: 1968–1972, edited by David Stradling

Reel Nature: America's Romance with Wildlife on Film, by Gregg Mitman

DDT, Silent Spring, and the Rise of Environmentalism, edited by Thomas R. Dunlap

Conservation in the Progressive Era: Classic Texts, edited by David Stradling

Man and Nature: Or, Physical Geography as Modified by Human Action, by George Perkins Marsh

A Symbol of Wilderness: Echo Park and the American Conservation Movement, by Mark W. T. Harvey

Tutira: The Story of a New Zealand Sheep Station, by Herbert Guthrie-Smith

Mountain Gloom and Mountain Glory: The Development of the Aesthetics of the Infinite, by Marjorie Hope Nicolson

The Great Columbia Plain: A Historical Geography, 1805–1910, by Donald W. Meinig

Weyerhaeuser Environmental Classics is a subseries within Weyerhaeuser Environmental Books, under the general editorship of Paul S. Sutter. A complete listing of the series appears at the end of this book.

MAKING CLIMATE CHANGE HISTORY

*Primary Sources from
Global Warming's Past*

EDITED BY
JOSHUA P. HOWE

UNIVERSITY OF WASHINGTON PRESS
Seattle and London

Making Climate Change History is published with the assistance of a grant from the Weyerhaeuser Environmental Books Endowment, established by the Weyerhaeuser Company Foundation, members of the Weyerhaeuser family, and Janet and Jack Creighton.

Copyright © 2017 by the University of Washington Press
Printed and bound in the United States of America
Design by Thomas Eykemans
Composed in OFL Sorts Mill Goudy, typeface designed by Barry Schwartz
21 20 19 18 17   5 4 3 2 1

All rights reserved. No part of this publication may be reproduced or transmitted in any form or by any means, electronic or mechanical, including photocopy, recording, or any information storage or retrieval system, without permission in writing from the publisher.

University of Washington Press
www.washington.edu/uwpress

Library of Congress Cataloging-in-Publication Data on file

ISBN (hardcover): 978-0-295-74138-3
ISBN (paperback): 978-0-295-74139-0
ISBN (ebook): 978-0-295-74140-6

The paper used in this publication is acid-free and meets the minimum requirements of American National Standard for Information Sciences—Permanence of Paper for Printed Library Materials, ANSI Z39.48–1984. ∞

CONTENTS

Foreword: Climate Change and the Uses of History,
    by Paul S. Sutter . . . . *xi*
Acknowledgments. . . . . *xv*

Introduction: Making Climate Change History 3

PART 1 THE SCIENTIFIC "PREHISTORY" OF GLOBAL WARMING 21

Joseph Fourier, "General Remarks on the Temperatures
of the Globe and the Planetary Spaces" (1824) 27

John Tyndall, "The Bakerian Lecture: On the Absorption
and Radiation of Heat by Gases and Vapours, and on
the Physical Connexion of Radiation, Absorption, and
Conduction" (1861). 32

Svante Arrhenius, "On the Influence of Carbonic Acid
in the Air upon the Temperature of the Ground" (1896) 39

G. S. Callendar, "The Artificial Production of Carbon
Dioxide and Its Influence on Temperature" (1938) 45

PART 2 THE COLD WAR ROOTS OF GLOBAL WARMING. 49

Roger Revelle and Hans E. Suess, "Carbon Dioxide Exchange
between Atmosphere and Ocean and the Question of an
Increase of Atmospheric CO_2 during the Past Decades" (1957) 55

Roger Revelle, Testimony before the House Committee on
Appropriations, February 8, 1956 60

Roger Revelle, Testimony before the House Committee
on Appropriations, May 1, 1957. 64

Howard T. Orville, "The Impact of Weather Control
on the Cold War" (1958). 70

National Science Foundation, *Preliminary Plans for
a National Center for Atmospheric Research* (1959) 77

PART 3 MAKING GLOBAL WARMING GREEN 85

The Conservation Foundation, *Implications of Rising Carbon
Dioxide Content of the Atmosphere* (1963) 91

President's Science Advisory Committee, *Restoring
the Quality of Our Environment* (1965) 96

Donella H. Meadows, Dennis L. Meadows, Jørgen Randers,
and William W. Behrens III, *The Limits to Growth* (1972) 103

Study of Man's Impact on Climate, *Inadvertent Climate
Modification* (1971) 108

The Sierra Club, "International Committee Questionnaire—
Five Year Plan" (1976) 115

Michael McCloskey, "Criteria for International Campaigns"
(1982). 121

National Climate Program Act of 1978. 123

American Association for the Advancement of Science,
Advisory Group on Climate Meeting, May 26, 1978 128

David Slade, "Action Flow, U.S. Carbon Dioxide Research
and Assessment Program" (1979) 132

David Slade, Letter to David Burns (1980). 134

Al Gore, Testimony before the House Committee on Science
and Technology, July 31, 1981 136

Rafe Pomerance, testimony before the House Committee
on Science and Technology, February 24, 1984 141

PART 4 CLIMATE CHANGE AS CONTROVERSY 145

U.S. Central Intelligence Agency, "A Study of Climatological
Research as It Pertains to Intelligence Problems" (1974) 151

S. I. Rasool and S. H. Schneider, "Atmospheric Carbon
Dioxide and Aerosols: Effects of Large Increases on Global
Climate" (1971) 161

Reid Bryson, "A Perspective on Climate Change" (1974) 164

Stephen H. Schneider, *The Genesis Strategy* (1976) 166

    Helmut E. Landsberg, "Review: *The Genesis Strategy—
    Climate and Global Survival*" (1976) 169

    Stephen H. Schneider and Helmut E. Landsberg, "Forum"
    (1977) 172

National Academy of Sciences, "Carbon Dioxide
and Climate" (1979) 175

National Academy of Sciences, "Changing Climate" (1983) 180

Environmental Protection Agency, *Can We Delay
a Greenhouse Warming?* (1983) 189

    New York Times, "How to Live in a Greenhouse" (1983) 193

R. P. Turco, O. B. Toon, T. P. Ackerman, J. B. Pollack, and
Carl Sagan, "Nuclear Winter" (1983) 195

Carl Sagan, "Nuclear War and Climatic Catastrophe" (1983) 197

S. Fred Singer (1985), "On a 'Nuclear Winter'" (1983) 201

Starley L. Thompson and Stephen H. Schneider, "Nuclear Winter Reappraised" (1986) 203

James Hansen, Testimony before the Senate Committee on Energy and Natural Resources, June 23, 1988 206

PART 5 CLIMATE CHANGE GOVERNANCE 209

Intergovernmental Panel on Climate Change, First Assessment Report (1990). 215

World Commission on Environment and Development, *Our Common Future* (The Brundtland Report) (1987) 220

United Nations, Rio Declaration on Environment and Development (1992) 224

United Nations Framework Convention on Climate Change (UNFCCC) (1992). 229

C. Boyden Gray and David B. Rivkin Jr., "A 'No Regrets' Environmental Policy" (1991). 238

Al Gore and Mitch McConnell, Testimony before the Senate Committee on Foreign Relations, September 18, 1992. 241

Intergovernmental Panel on Climate Change, Second Assessment Report (1996). 247

The Kyoto Protocol to the United Nations Framework Convention on Climate Change (1997). 255

The Byrd-Hagel Resolution (1997). 264

PART 6 THE PAST, THE PRESENT, AND THE FUTURE 269

Bill McKibben, *The End of Nature* (1989). 277

Paul J. Crutzen and Eugene F. Stoermer, "The Anthropocene" (2000) 282

Michael Shellenberger and Ted Nordhaus, "The Death of Environmentalism" (2004). 287

Nicholas Stern, "Stern Review on the Economics of Climate Change" (2006). 294

    William D. Nordhaus, "A Review of the *Stern Review on the Economics of Climate Change*" (2007). 302

Massachusetts v. Environmental Protection Agency (2007) 307

Pope Francis, *Laudato Si': On Care for Our Common Home* (2016). 321

Index. 333

    Historicizing Data color plates appear after page 208

FOREWORD

Climate Change and the Uses of History

PAUL S. SUTTER

What is the use of history in the face of potentially catastrophic climate change? What wisdom might the study of the past offer us in the present to help us contend with a future of profound environmental disruption? These are critical questions, and yet they have been difficult to raise in conspicuous ways because so much public attention—and contention—has focused on the science of global warming and the case scientists have made, now indisputably, that humans have become agents of biogeochemical transformation on a planetary scale. This focus has been understandable. A problem of such scale, complexity, and obscurity has demanded the kinds of careful empirical and conceptual work that scientists have brought to it. Their work has been both remarkable and indispensable. But human-induced climate change is more than a scientific problem, and to fully contend with it we need other skills and habits of mind, including those cultivated by the discipline of history. While climate change may be ushering us into uncharted historical territory, the past is still the richest imaginative resource we have for understanding and shaping our common future.

In the face of climate change, history's uses are many. We need careful historical explanations of how and why humans turned to fossil fuels, unleashing huge stores of fossil carbon and other greenhouse gases into the atmosphere and pushing the global climate system into unprecedented territory. More than that, the discipline of history is essential to understanding the forces of economic, social, political, diplomatic, and cultural power that have driven us into a world of human-induced climate change, and to appreciating how effective solutions must contend with those power dynamics and the inequalities and injustices they have spawned. Climate change is a human problem, but it is not the product of an abstract and universal humanity. Instead, it is a result of global concentrations of power that historians are well equipped to help us understand. If the most important solution to the climate crisis is a rapid transition to carbon-neutral energy sources, a sophisticated understanding of past energy transitions and their broad societal implications will be vital to achieving such a goal. The siren call of simple technical fixes is a strong one, but the study of the past compels us to resist its seductiveness. We also have much to learn from the burgeoning field of climate history and from the impacts of past climate disruptions, from the Little Ice Age and its shaping effects on the early modern world to punctuated events such as the eruption of Mount Tambora in 1815 and the myriad global crises that followed in its wake. Such historical case studies enrich our thinking about how climate change may disrupt human societies in the future. Global climate change is a fundamentally historical problem, a problem of changes in human behavior over time, and we need skilled and well-trained historical thinkers to help us explain and navigate this new world we have created.

One of the most perplexing historical questions we face in relation to climate change is why, despite an overwhelming scientific consensus on its human causes that is decades old, the world community has been so unresponsive to the emerging crisis. This is the terrain of Joshua Howe's superb new reader, *Making Climate Change History*. Howe insists that we must think deeply and critically about how we have come to understand the science of climate change, and how experts have tried to translate their knowledge into public discourse and meaningful public policy. If one were to quickly browse through the table of contents without reading Howe's incisive introduction, one might conclude that this reader is something like a greatest hits collection, a compendium of the most important documents in the incremental development of our current understanding of the climate change problem. That is not an entirely inappropriate way of reading Howe's selections, for all are documents that have, as he puts it, "made history"—they are critical building blocks in the larger edifice that is contemporary climate change science and policy. But Howe also warns us against what he calls the "presentist paradox": that if we read these collected documents merely as harbingers of our current conundrum, as a popcorn path to our present state of climate change consciousness, we will miss their historical specificity and the contextual forces and influences that shaped them. Howe's deeper agenda here, then, is to present these documents as the raw materials from which readers can make their own histories of expert knowledge and public discourse on the issue of climate change, and come to their own understandings about the rocky relationship between climate science and public policy.

As a historian of science, Howe also invites us to use these documents to rethink the very narrative of scientific progress. This is not because climate scientists have not made "progress" on understanding how our climate system works. On the contrary, one of the great historical lessons that this brief volume illustrates is how, over a relatively short period of time, scientists have come to understand global climatic and atmospheric systems, and human impacts upon them, with remarkable scope and specificity. The point, rather, is that such narratives of intellectual progress tend to assume a too simplistic historical relationship between ideas and action. They rip science from its historical context, and they suggest that scientific truth should unfailingly yield rational political progress. But such narratives confound us precisely when scientific truth does not lead to a clear policy path and a broad public commitment to follow it. Understanding science in context, then, is not just an academic exercise. It is the essence of the problem of how we turn what experts now know about anthropogenic climate change into meaningful and effective policies to combat its effects.

Part of the genius of *Making Climate Change History* is its organization and structure. Again, a quick look at the table of contents might suggest a conventional chronological approach, but Howe has more ambitious goals. Howe begins the volume with a chapter on the foundational documents in atmospheric science, partly to give his readers a sense of the origins of modern climate change concern. But he also offers these documents as what he calls "wormholes into other historical worlds," as sources meant to tell us as much about their moments of production as about the scientific present they have shaped. This first chapter is also an opportunity for Howe to provide some guidance on how to read scientific papers as historical documents. Subsequent chapters follow a similar template: they introduce a specific context within which climate science and policy developed, and they focus on different types of documents with which historians work. So there are chapters on the Cold War context of climate science, the meeting of climate science and the early environmental movement, the politicization of climate science in the United States, the rise of international governance around climate change, and how climate change has recently entered popular discourse. In these chapters, Howe gently instructs us on how to read sources such as congressional testimony, scientific gray literature, government reports, newspaper editorials, international diplomatic frameworks, popular writing, and even visual images of scientific data such as the Keeling Curve and hockey stick graph, which have become the most recognizable vehicles for communicating climate science to the public. In the process, Howe makes a case for historical literacy as a critical skill in what we might call modern climate citizenship.

Making Climate Change History not only introduces readers to essential documents in the history of climate science and policy; it invites readers to work with them as a historian would. Howe is the perfect tour guide for such an exploration, for his recent book, *Behind the Curve: Science and the Politics of Global Warming* (also a Weyerhaeuser Environmental Book), is one of the most important studies we have of why clear climate science has not always resulted in coherent climate action. In *Making Climate Change History*, Howe effectively takes us behind the scenes of his previous scholarship and provides his readers with a primer on how to make sense of the interactions between experts working "at the boundaries of their expertise" and popular action and reaction to their ideas. As such, this is a volume whose value extends to anyone who is interested in the historical relationship between science and policy. It is a volume committed to the proposition that thinking historically matters in the face of an uncertain future. It is a volume that is dedicated to making climate change history.

ACKNOWLEDGMENTS

I had a lot of help in every phase of editing this collection. Bill Cronon and Marianne Keddington-Lang at the University of Washington Press proposed that I put the collection together just as my own history of global warming, Behind the Curve, went to press, and Paul Sutter and Regan Huff coaxed me to complete it once I realized just how much work it would be. Their patience, feedback, and encouragement made the book possible.

Adam Pease, Zeus Smith, and Sara Keleman did a good deal of the transcription work on this project—the heavy lifting—and along the way they provided the kind of valuable, unguarded insight that only bright undergraduates can. Along with the students in my 2015 American Earth course at Reed College, who workshopped the collection with me, they deserve special thanks and a good deal of credit for producing this work. Thank you to the Department of History and the Environmental Studies Program at Reed, and to the institution more broadly, for their generous and continued support. Teaching at Reed has helped shape the way I think about the value of history to environmental issues, and it defines how I teach history to others.

David Stradling, Jim Feldman, and Chris Wells provided excellent suggestions and cautionary advice on the project in its early phases; Fredrik Jonsson and the University of Chicago faculty and graduate students who participated in the 2015 Neubauer Collegium offered valuable feedback on portions of the work as it took a clearer form. I would also like to thank Bob Wilson for his thoughtful and detailed review of the essays and included materials. Finally, as with everything I write, my approach to this work, and to history more generally, owes its best qualities to Richard White.

Perhaps most importantly, this book is the product of the hard thinking and tireless efforts of the people whose work on the science and politics of climate change I have excerpted here. Thank you for helping make climate change history.

MAKING CLIMATE CHANGE HISTORY

INTRODUCTION

Making Climate Change History

This is a book about two things together: history and climate change. It is not meant to provide a comprehensive survey of either; rather, the book is an attempt to use documents from global warming's past as a way to think about doing history, while at the same time using history as a way to approach the politics of climate change from a new perspective. The objective of the book is to provide the material for readers to pose critical and creative questions regarding these documents as a way to help them understand climate change in historical context. The argument of the book is that when we look at problems related to climate change, thinking historically matters.

As the title suggests, this is a book about making climate change history in a few related senses. First, as a curated set of documents—a "reader," as we call them—the book presents written evidence from the people, communities, and institutions that first articulated anthropogenic global warming as a problem, brought that problem to the attention of politicians and the general public, and attempted to establish a variety of forms of governance in order to address the problem at local, national, and international scales. Each document represents a part of the scientific and political history of climate change at a given moment in the twentieth century. The collection includes some documents from the scientific literature, but I focus primarily on those pieces of evidence that show how scientists, politicians, and activists worked to transform expert knowledge into a broader public discourse. The authors of these documents were primarily experts in technical fields like atmospheric chemistry, physical oceanography, political science, or environmental law, but the documents themselves reflect their efforts to deal with an emerging global problem that often pushed them beyond the boundaries of those fields. In their own context, these documents shaped and drove a broader conversation about anthropogenic global warming in important ways, and they were important enough in their own time to significantly influence the way we talk about climate change today. In common speech, we often describe agents of social, cultural, or political change as the people or things that "make history." In this sense, these are not just documents from global warming's past; they are the documents that "made global warming history."

There is another sense in which this collection addresses the making of climate change history, however, and it presents something of a challenge to the whole idea of "making history" in the first, colloquial phrase. For many historians, the common understanding of history as a series of momentous events from our collective past masks a key difference between the raw materials of historical evidence and the types of stories that both professional historians and members of society more broadly tell about that evidence. At its core, writing or "doing" history is a creative act. The stuff of the past does not and cannot speak for itself; it requires an active interlocutor, a narrative tour guide that puts pieces of the stuff of the past—evidence—together in ways that make it make sense within its own context. The chronicle of history—the nearly infinite lists of causally disconnected facts from the past—in the end bounds our interpretation of the past, but does very little to make it either meaningful or useful.[1] The chronicle of the past does not teach us lessons; we learn lessons by working with that chronicle to make it meaningful. Insofar as historians build plausible, compelling, and insightful stories from the body of evidence that constitutes the chronicle, historians themselves make history as much as their subjects do.

To me, as a historian of the science and politics of climate change, there is a funny thing about thinking about history this way. I have written almost exactly these words about the lessons from the science of climate change. Much like the stuff of the past, evidence of anthropogenic climate change—raw temperature and weather data, advanced climate models, and even weather and climate events—becomes meaningful only through the work of interlocutors.[2] Most often those interlocutors are scientists, who put pieces of evidence in conversation with each other and present the results in ways that make scientific, political, economic, and moral sense. Insofar as climate scientists seek to make meaning out of their research, making climate change meaningful and making history represent significantly similar cultural and intellectual projects. They require piecing evidence together in reliable and culturally relevant ways that teach us lessons—lessons about the past, about the present, and though many historians shy away from the idea, also about the future.

This is the second meaning of the title of this book. *Making Climate Change History* is an invitation to practice the historian's creative craft. The book presents selections from the chronicle of climate change history representing a variety of different types of historical documents alongside a series of essays that at once help to put these pieces of evidence in their specific historical contexts and offer guidance in making historical sense of these and other similar documents. Again, this is not a comprehensive guide to either historical methodology or the history of climate change. But the diversity of the documents—scientific papers, liminal scientific literature, congressional hearings, newspaper articles, internal organizational correspondences, documents produced by international legal and political organizations, and other materials produced by experts at the confluence of science, policy, and public knowledge—presents an opportunity to explore a variety of historical approaches. By introducing questions and specific strategies for addressing the particular types of documents produced by the history of climate change, the essays offer a set of tools for interpreting this discourse on its own terms. *Making Climate Change History* thus presents both a curated chronicle of global warming's past and a loose how-to guide for making sense of and telling stories about that chronicle.

Finally, there is an aspirational meaning to *Making Climate Change History*. There is, of course, a vanishingly small probability that a document reader will somehow put an end to anthropogenic warming; in fact, both the geophysical and historical evidence suggest that the problem is only getting worse. That said, efforts to mitigate and adapt to climate change continue to grow in scale and change in kind, and their relative success depends in part on recognizing how climate change discourse has changed in the past and continues to change in the present. Read carefully and on their own terms, the documents in this collection demonstrate how the stakes of conversations about climate change—about its causes and consequences—have become increasingly clear over time. As the stakes have changed, so too have the social and political responses to climate change changed, and understanding what has changed in the conversation and what has stayed the same—that is, understanding the history of the issue—is important in building just, equitable, and effective strategies for dealing with the climatic challenges of the future.

In practical terms, engaging the history of climate change contributes to contemporary climate change discourse in two important ways. The first is, in a way, strategic. As the documents here reveal, to varying degrees, scientists, environmentalists, and some politicians have understood climate change as an important threat to species, ecosystems, and human well-being for more than half a century. And yet not only have the concentrations of CO_2 and other greenhouse gases continued to rise during that period, the rate of that rise has actually increased. Society certainly knows and cares more about climate change than it did in the 1950s, but while better information and greater concern have changed the discussion, effective solutions to the problems of climate change have been elusive. Why? What has gone well in the scientific and political history of climate change, and what, from the perspective of a concerned citizen in the twenty-first century, has gone wrong? *Making Climate Change History* provides an opportunity to play back the tape in order to understand the patterns of scientific, political, and public engagement that have informed—and continue to inform—the climate change conversation since its inception.

There is also a second, perhaps more profound value to thinking historically about climate change. As a "future tense" problem, climate change threatens to upset social, political, economic, and ecological structures, but assessing those threats requires a careful consideration of what we have valued about those structures in the past and why we care about them in the present. That consideration is fundamentally historical. Historical thinking—that is, an evaluation of evidence on its own terms, predicated on a nuanced understanding of context and contingency—provides a critical perspective from which to evaluate how climate change may make the world of the future different from the worlds of the past and present. By introducing some of the tools that historians use to approach their evidence, *Making Climate Change History* provides a starting point for developing and honing this kind of historical sensibility. It is a sensibility often overshadowed by the urgency of the issue of climate change, but one that I believe is essential in thinking about equitable, just, and effective responses to climate change in the future.

THE FAMILIAR AND THE STRANGE

I have divided the documents in this collection into six parts and one color insert dedicated to scientific images. The organization is primarily chronological, but the documents also reveal different sets of themes and sometimes require different methodological approaches, reflecting the increasing breadth and heightened concern of conversations about climate change as those conversations changed over time. Each part begins with an essay that focuses on selected documents in that part to explore methodological questions about a particular type of document, while also providing some context for the part as a whole. Because the introductory essays don't tackle every document in each part, I have also provided brief document headers designed to provide enough context for readers to take their exploration beyond the introductory essays—and on to further reading, highlighted in the notes—if they so choose. Taken together, the essays, headers, and documents provide a framework for making sense of historical evidence from global warming's past.

Making Climate Change History begins with a set of documents from a period in the nineteenth and early twentieth centuries when European and American scientists began to articulate some of the basic concepts that underpin modern ideas about how CO_2 works to warm the earth. As early as the seventeenth century, Western scientists had identified the nature of heat as one of the fundamental questions of the physical world. Following the advent of the steam engine and the subsequent rapid industrialization of Europe, questions about heat began to overlap with a larger set of questions about the equivalences among motion, heat, and work. As nineteenth-century scientists began to investigate the nature of heat and the developing laws of thermodynamics using new mathematical and experimental approaches, they landed on conclusions about the way the atmosphere absorbs and reradiates heat that begin to look familiar to modern readers interested in climate change.

Scientific papers from the nineteenth century outlining fundamental concepts like the greenhouse effect can be useful and important for a variety of reasons. For historians of science, these papers lie at the beginning of a trail of citations that can help in reconstructing how later climate scientists came to further insights about the earth's atmospheric and climatic systems. Historical scientific papers also reinforce the credibility of modern climate science in the face of cynical attacks by climate change skeptics and deniers by demonstrating the arduous processes by which knowledge about the earth's climate has been won.[3] By showing the long-standing scientific interest in climate change and the slow accretion of knowledge about how the climate system works, historians help undercut some of the specious claims lobbed at concerned scientists: that they have succumbed to a faddish doomsdayism or, worse, a form of base opportunism. The standard narrative of the nineteenth- and early twentieth-century history of global warming draws a neat narrative line of scientific discovery from Jean-Baptiste Joseph Fourier's studies of terrestrial temperatures in the 1820s to the current Intergovernmental Panel on Climate Change (IPCC) consensus on the science of climate change, and in doing so it lends the modern institution the additional weight of history.[4] The documents in a collection like this serve as a record of the conceptual, scientific, and technical developments that have enabled experts to comprehend the various pieces of this problem over time. If the representative international structure of the IPCC helps to ensure the accuracy and credibility of its reports in the here and now, its engagement with two hundred years of formative scientific inquiry only further underscores its rigor.

That standard narrative is both romantic and rhetorically useful, but it also highlights one of the key challenges in making climate change history. I call this problem the presentist paradox. It works like this. Presentism in history is a flaw wherein the social and political concerns of the present that provide the motivation for tackling a certain historical subject matter overdetermine interpretations of the past. It is a form of confirmation bias. Presentist histories trend toward the teleological; they focus on the way past trends and events have led to the familiar textures of the present. They err when that focus on the present comes at the expense of the strangeness of the specific contexts in which events in the past took place. The strangeness of the past and the attention to the contexts in which strange things happen help make history a valuable way to understand the structures and contingencies that shape our world. Without that strangeness, histories become narrow "just-so" stories that mostly tell us things we already know and reveal very little about human experience.

The paradox comes in choosing the material that constitutes climate change history. We care about climate change history primarily because we care about the modern problem of climate change; climate change history thus almost by necessity has a presentist flavor to it. Even where careful historians have investigated particular moments in climate change history on their own terms, the moments we choose tend to reflect our collective knowledge of where those moments fit within a larger trajectory that only ends in the present.[5] While good history seeks to interpret documents in their contemporary contexts, often the only way to identify those documents that count as evidence in global warming's past is to look for where those documents engage with familiar modern ideas about global warming. And that tends to distort the past.

Not all histories written from the present succumb to the teleological temptations of presentism, however, and even when it does come with heavy doses of teleology, not all presentist history is bad history. Dealing with the presentist paradox represents a key theme of this book. To begin, part 1 introduces strategies for approaching scientific documents, which have come to constitute the standard narrative of climate change history. These strategies balance a present-minded focus on climate change with a historian's commitment to exploring documents in their specific historical contexts and on their own terms. How, for example, do scientific documents differ from other sorts of historical sources? How do you balance your particular interest in the piece you are looking at—in this case, historical references to climate or climate change—against the larger focus of the scientists themselves? How do scientific papers and scientific practice in, say, 1863 look similar to practices and publications of science today, and how are they different? Selections from four classic documents—Joseph Fourier's 1822 investigation of the temperatures of the earth and planetary spaces; John Tyndall's 1861 paper on the radiative properties of gases, which identified CO_2 and water vapor as greenhouse gases; Svante Arrhenius's 1896 mathematical model of atmospheric CO_2 doubling; and G. S. Callendar's 1938 paper on anthropogenic climate change—highlight both the striking similarities and important differences between early scientific discourse on global warming and our modern concerns.

The presentist paradox comes in many flavors. Whereas part 1 requires a renewed attention to scientific documents that have been interpreted in a problematically presentist light, part 2 highlights the ways in which, even very early on in the history of global warming, published scientific papers represent only one of many types of historically situated sources from which to construct global warming's story. The documents from part 2 not only highlight important scientific advances related to climate change; they also revolve around the integration of climate science into the Cold War research system of the late 1950s and early 1960s. For example, Roger Revelle's famous 1957 paper in the journal *Tellus* demonstrates the scientific community's interest in CO_2 during this period, but the accompanying congressional testimony also demonstrates how Revelle's scientific interest both informed and responded to a range of political, scientific, and cultural issues associated with the Cold War of the late 1950s.

Few periods of American history require more attention to this public face of science than the first decades of the Cold War. The Cold War helped create a sense of vulnerability to both intentional and unintentional anthropogenic disasters that set the stage for a broader public and political interest in the issue of climate change.[6] A keen interest in the military value of intentionally modifying nature's geophysical systems to defeat a Soviet enemy touched off fears about inadvertently modifying those same systems in ways that might lead to environmental catastrophe. Popular concerns about nuclear energy in particular highlighted the extent to which humans in the postwar world had the power to change nature on a grand—even global—scale. Meanwhile, in part as a result of these fears, the Cold War also fostered an era of international scientific competition that justified the expansion of a basic scientific research system fundamental to the development of climate science in the 1960s and 1970s. Scientific publications like Revelle and Suess's 1957 *Tellus* article reveal some of the hopes and anxieties that attended these developments. In the halls of Congress, where Revelle and his colleagues worked to secure government support for their research and their institutions, scientists engaged with these hopes and fears in ways they seldom did in scientific publications.

The introductory essay highlights the challenges and opportunities of writing history from congressional testimony, where a range of actors—both scientists and others—speak in highly politicized rhetoric to support a range of overlapping and sometimes conflicting personal, political, and institutional interests. Who was enlisted to speak for science in the public sphere, and how did their voices change when they did so? What was at stake when scientists appeared before Congress in the 1950s, and for whom? What cues can you look for between the lines of testimony to access the interests and motivations of the people in the room? And more generally, how do scientists' public statements reflect their scientific, institutional, and personal political objectives?

The Cold War helped shape climate change discourse throughout the second half of the twentieth century in significant and sometimes unpredictable ways, but it was not the only formative influence on the science and politics of climate change. The integration of climate science into the Cold War research system overlapped with the rise of the modern American environmental movement, and between Revelle's 1957 *Tellus* paper and the advent of the Intergovernmental Panel on Climate Change in the late 1980s, climate change moved from the fringes of Cold War research to the center of American and international environmental politics. That transition was episodic, contingent, and often highly problematic. Part 3 provides a series of documents that demonstrate the challenges of making global warming green in a variety of different historical, institutional, and political contexts. Beginning with Charles David Keeling's 1963 report for the Conservation Foundation, in which he defined CO_2 as a form of pollution, the documents highlight the difficulty of making a highly technical, largely intangible, long-term global problem mean something to constituencies in environmental and scientific organizations, in geopolitical negotiations, and among the larger American public.

Perhaps no set of materials captures the challenges of making global warming green as well as those related to the Sierra Club International Program's strategic planning efforts in the late 1970s. In 1976, the club designed a questionnaire on international environmental priorities to send to each of its chapters as it began to design a five-year plan for environmental initiatives abroad. Though international program director Patricia Scharlin saw issues of the global commons—including climate change specifically—as a unifying set of concerns that might tie together the club's other international efforts, the club's local constituencies made climate change a low priority, if they mentioned it at all.

The Sierra Club surveys also present a different type of document in the history of climate change, highlighting the importance of understanding institutions in approaching historical materials. Part 3 suggests some ways institutional structures help shape history and can thus guide our reading of archival documents. How, for example, do the Sierra Club surveys—documents from the archives of a grassroots environmental organization—differ in content and form from documents produced by the American Association for the Advancement of Science and the Department of Energy (DoE) in their effort to incorporate climate change into DoE planning for the 1980s? What types of information are most important to each institution, and how do the documents treat that information? How do the authors' roles in their organizations—and their organizational missions—show up in the texts?

At the same time environmentalists began to think about how to incorporate the problem of climate change more meaningfully into environmental advocacy, climate change discourse began to take on new political dimensions that made the issue more controversial, both within the scientific community and outside of it. Today, we tend to think of climate change as a divisive partisan issue, and for good reasons. Al Gore, a two-term vice president and Democratic presidential candidate in 2000, championed climate change policy before and after his exit from electoral politics, while some of global warming's most vocal skeptics have represented the Republican party in the House, the Senate, and the White House. In 2015, the Pew Research Center found that only 20 percent of Republicans polled considered climate change a "very serious problem" compared to 68 percent of Democrats, and about 50 percent of Republicans supported curbing CO_2 emissions compared to 82 percent of Democrats.[7] Media coverage has tended to magnify this partisanship. But while climate change has certainly always been political, the politics of climate change have not always played out in the same places or along the same ideological lines that they do today. Here again, thinking too much about what looks familiar and not enough about what looks strange can blind us to some of the important contours of climate change history.

Part 4 introduces documents from the 1970s, 1980s, and early 1990s that highlight a change in the tenor of climate change discourse as the politics of climate change deepened and polarized during that period. The documents revolve around three episodes: the debate between proponents of a global cooling hypothesis and a global warming hypothesis in the early 1970s; the publication of conflicting reports by the National Academy of Sciences and the EPA in 1979 and 1983; and the debates over nuclear winter that took place in the first half of the 1980s. On the surface, each of these debates looks like a product of basic scientific disagreement, but a more careful analysis reveals that divisions in each case arose as much over how protagonists understood the interfaces between science and policy as they did over more specific technical disputes within the science itself. In addition, each conflict reached beyond the pages of scientific journals and spilled over into public view via popular magazines, press releases, government assessments, and newspaper reports. These liminal scientific sources—documents produced by scientists but published outside of the peer-review structures of normal scientific publication—hold the key for understanding the polarization of what was eventually dubbed the climate change debate, despite the overwhelming evidence supporting the theory of anthropogenic warming. Because these liminal documents come in many shapes and sizes, however, they require a particular attention to form and context. What kinds of differences might you expect, for example, between a dispute presented in a book review in *Science* and a disagreement between scientists in the popular press? How might a collaborative scientific report enable or constrain the way scientists present the seriousness of the ramifications of climate change? And how do scientists maintain their claims to authority and objectivity when they choose to express their views in these liminal sources? Part 4 provides some ideas on what types of questions historians can ask to make sense of these documents.

The scientific controversies that arose in the 1980s gained new importance in the 1990s, when scientists, environmentalists, and some politicians began to marry climate change discourse to international programs of environmental governance. Chapter 5 documents some of the key debates surrounding the confluence of climate change and international environmental governance between 1985 and 1998. Moving beyond the confines of U.S. institutions that characterize the other sections of this book, part 5 focuses on the structures of documents and treaties created by a variety of United Nations–affiliated organizations. Beginning with the introduction of the slippery concept of sustainable development in the World Commission on Environment and Development's report, *Our Common Future*, the section pays particular attention to how different types of international scientific and diplomatic institutions produce documents and how those processes of production overlap. In particular, the documents demonstrate how the contested concept of sustainable development helped tie together scientific and political processes in both national and international forums.

The most important scientific organization involved in the debate over climate change governance was the Intergovernmental Panel on Climate Change (IPCC). Founded in 1988, the IPCC emerged as a boundary organization meant to mediate between the science and politics of climate change.[8] How did the organization's status as a boundary organization impact the drafting process of IPCC assessment reports, and how did that drafting process change with each new assessment? Scientific meetings, World Bank policy documents, and Senate hearings reveal how politicians and policymakers understood the IPCC and related United Nations Framework Convention on Climate Change (UNFCCC) in terms of a changing late– and post–Cold War geopolitical context. How do the texts of IPCC documents differ from those of the UNFCCC and the World Bank, and what overlaps reveal their common origins? In tracking the relationships among policy documents in these changing contexts, this section offers a strategy for making sense of international climate change politics in the 1990s.

If the 1990s saw the establishment of a more robust interface between science and policy, it also saw an explosion of new forms of climate change discourse that vastly expanded popular, academic, and political interest in the subject matter. Whereas scientific experts writing about climate change between the 1950s and 1990s focused largely on explaining the technical details of climate change for policymakers, since the late 1980s experts—many of them not scientists—from other fields have begun to focus on what climate change might mean for them. Beginning with Bill McKibben's *The End of Nature*—a popular 1989 exposé exploring what climate change meant for humans' relationship to nature and wilderness that defined climate change for a generation of Americans—part 6 addresses some of the ways journalists, judges, religious leaders, economists, and political strategists have made meaning out of climate change since McKibben's groundbreaking work.

More than in any other section, the self-conscious meaning-making that appears in part 6 underscores the role of history in making sense of global warming. Each of the documents addresses a relatively simple question: "What has climate change changed?" In doing so, they draw a line between the past, the present, and the future intended to highlight what has changed and what has remained the same. *The End of Nature*, for example, frames the altered world of the present and future against a simpler, purer past in order to highlight the profound rupture caused by human intervention in the global environment. In 2003, political strategists Michael Schellenberger and Ted Nordhaus asked how climate change had changed the political context of environmental activism, and four years later the Supreme Court considered what climate change meant for issues of standing and congressional intent in a 5–4 decision on *Massachusetts v. EPA*. Most recently, Pope Francis addressed the meaning of climate change for Catholic doctrine in his encyclical *Laudato Si', On Care for Our Common Home*. Each of these documents makes a case for the meaning of climate change based on an interpretation of historical precedent, underscored in some cases by elaborate formal notation (much of which I have included in this part). And yet, none of these authors self-consciously identify as historians, nor are they writing histories. So how do we assess the way these authors use the past to make meaning of climate change? How has the appeal to historical precedent helped the pope, the Supreme Court, or British Treasury economist Nicholas Stern make meaning out of climate change in their particular contexts?

Finally, in addition to its main chapters, this book contains a color insert in order to tackle another type of historical document altogether, one without which no study of the climate change history could be complete: the data image. Throughout the last sixty years of climate change discourse, data images have played a key role in communicating the causes and potential impacts of anthropogenic warming. Largely the product of scientific exchange, data images also make an odd form of cultural artifact that conveys contextually specific information to culturally and historically specific communities in the same way a photograph or a piece of art might. How do we read data images as historical documents? Beginning with one of the simplest and most common data images in the history of climate change, the upward sloping oscillating line of measured atmospheric CO_2 from 1957 called the "Keeling Curve," the insert explores the ways in which data images—and challenges to them—can convey information with scientific, political, and broader social meaning. Here, the introductory essay is meant as food for thought, a sort of template for investigating the other images—including scenario graphs from successive IPCC reports and the famous "hockey stick" graph that became a centerpiece of controversy in 2009—in their own particular historical contexts.

HOW TO USE THIS BOOK

There are a variety of ways to use this book. One is to read it like any other book: start at the beginning and read until the end. Reading *Making Climate Change History* from start to finish provides an episodic and somewhat idiosyncratic narrative of the science and politics of climate change through the end of the twentieth century. The collection of documents is not comprehensive, nor do the essays provide a complete guide to reading historical documents, but the book nevertheless provides a relatively brief standalone primer in the history of one of the most important challenges of our time. As a group of documents in one place, the book also works as a reference, and alongside other histories of climate change—Spencer Weart's *The Discovery of Global Warming*, for example, or my own *Behind the Curve*—the collection offers a sort of behind-the-scenes documentary supplement, a companion to histories already made that provides a deeper understanding of the changing stakes of climate change throughout the twentieth century.[9]

Again, however, it is an incomplete reference, and it is worth pausing here for a moment to acknowledge some of what is missing and to explain some of what is here. First, this is neither a comprehensive collection of scientific papers nor a full collection of popular cultural texts or images. Of course, both categories of historical document are deeply important in understanding global warming's history, and I have incorporated some of each here.[10] The pages of peer-reviewed journals like

Science, the dust jackets of popular books like *The End of Nature*, the brochures of Greenpeace, and the advertisements of British Petroleum are all excellent places to look for the history of global warming. But to an impressive and perhaps unique degree in the context of modern social and environmental issues, the history of global warming has often played out most interestingly and meaningfully in other, less familiar forums, where its characters—mostly experts—operate at the boundaries of their expertise. This collection tracks those experts as they have attempted to make their expertise matter in National Academy of Science reports, in press releases, in congressional testimony, in international treaties, and in court.

Second, with the partial exception of part 5, this collection is heavily biased toward American sources. A global version of this reader—complete with the presentations of scientists and environmentalists to national governing bodies around the world—would be a very big book. To an extent, the American focus reflects both this practical limitation and my own interest in and experience with American sources. That said, there are also good historical reasons to focus on American sources in this collection. In the 1950s and 1960s, the United States emerged as a world leader in atmospheric science, and its institutions began to attract many of the most prominent scientists dealing with climate change either as visitors or as permanent residents. Until the late 1980s—and with a few important exceptions like the University of East Anglia and the British Meteorological Office—most of the major centers for climate change research were housed in the United States. So, too, did domestic U.S. politics serve as an important international touchstone for climate change politics elsewhere, particularly when it came to the development of international organizations and treaties like the IPCC and UNFCCC. Often, American politicians and political institutions have obstructed international negotiations on climate change more than they have facilitated them, but few aspects of international climate change politics evade the imprint of interactions among experts, advocates, and politicians in the United States. That trend continues, and understanding the American experience with global warming remains essential to making climate change history at any scale.

Finally, the idiosyncrasies of this book are also opportunities that speak to another, perhaps more fruitful way to use this book. At its best, *Making Climate Change History* is a series of starting points, wormholes into historical worlds both familiar and strange. Every piece presented here is an excerpt of a larger, publicly available document, and each one of those documents both tells a story and has a backstory. The notes point toward the documents; the essays provide some ideas on what to look for as you start to explore the milieu from which those documents emerged. Like the documents themselves, many of the backstories to the documents relate pretty obviously to the history of climate change, which, after all, is the guiding subject of this book. But the history of climate change is still an underdeveloped subfield, and many of these documents lead to stories that at first may seem unrelated or only tangentially related to the standard narrative of climate change, leaving important connections that still need to be made. There are important lessons in these yet undiscovered historical connections that can help shape the way we think not only about global warming's past, but also about its present and its future. They are lessons to be learned by making history, and they will not teach themselves.

NOTES

1. William Cronon, "A Place for Stories: Nature, History, and Narrative," *Journal of American History* 78, no. 4 (March 1992): 1347–76.
2. Joshua P. Howe, *Behind the Curve: Science and the Politics of Global Warming* (Seattle: University of Washington Press, 2014); Joshua P. Howe, "This Is Nature; This Is Un-Nature: Reading the Keeling Curve," *Environmental History* 20, no. 2 (April 2015): 286–93.
3. Naomi Oreskes and Erik Conway, *Merchants of Doubt: How a Handful of Scientists Obscured the Truth on Issues from Tobacco Smoke to Global Warming* (New York: Bloomsbury Press, 2010).
4. See David Archer, *The Warming Papers: The Scientific Foundation for the Climate Change Forecast* (Hoboken, NJ: Wiley, 2013); James R. Fleming, "Joseph Fourier, the 'Greenhouse Effect,' and the Quest for a Universal Theory of Terrestrial Temperatures," *Endeavour* 23, no. 2 (1999); James Rodger Fleming, *Historical Perspectives on Climate Change* (New York: Oxford University Press, 1998).
5. Fleming is among the most careful in this respect, as his 1998 title suggests.
6. Jacob Darwin Hamblin, *Arming Mother Nature: The Birth of Catastrophic Environmentalism* (Oxford: Oxford University Press, 2013).
7. "The U.S. Isn't the Only Nation with Big Partisan Divides on Climate Change," Pew Research Center, March 6, 2015, www.pewresearch.org

/fact-tank/2015/11/06/the-u-s-isnt-the-only-nation-with-big-partisan-divides-on-climate-change/.
8   For more on boundary organizations, see Stephen Bocking, *Nature's Experts: Science, Politics, and the Environment* (New Brunswick, NJ: Rutgers University Press, 2004).
9   Spencer R. Weart, *The Discovery of Global Warming*, rev. ed. (Cambridge, MA: Harvard University Press, 2008).
10 See Archer, *The Warming Papers*; Bill McKibben, ed., *The Global Warming Reader: A Century of Writing about Climate Change* (New York: Penguin, 2012).

PART I

THE SCIENTIFIC "PREHISTORY" OF GLOBAL WARMING

For historians of global warming, the early history of climate science presents a dilemma. On one hand, historians, scientists, and popular authors get significant traction out of a two-hundred-year story of investigations into the greenhouse effect that links an episodic series of scientific firsts into a scientific biography of global warming that reinforces the power of modern scientific consensus with the weight of historical time.[1] It is a useful story in that it demonstrates the impressive and often underappreciated longevity of the issue, and in the tradition of science, it gives credit where historians and scientists believe credit is due for new ideas that burgeoned into our modern understanding of CO_2-induced climate change.

And yet, this story is also deeply problematic. As a story of sequential firsts, the intellectual biography of global warming tends to presume a modern, twenty-first-century understanding of climate change that the story's nineteenth- and twentieth-century characters by and large did not have. The work of past scientists may seem prescient—we say that they were ahead of their time—but often that prescience is in part an artifact of our own retrospective viewpoint, and it can lead us to misunderstand our historical scientists' projects by failing to approach them on their own terms.[2]

Take, for example, John Tyndall's 1861 "On the Absorption and Radiation of Heat by Gases and Vapours." A prominent Irish physicist, mountaineer, and advocate of secular government widely famous for all three endeavors in the second half of the nineteenth century, Tyndall's twenty-first-century fame resides in his experimental verification of the greenhouse effect, the tendency of the atmosphere to absorb and reradiate energy from the sun in a way that helps to stabilize the temperature of the earth. Using an instrument called a ratio photospectrometer, Tyndall measured the opacity of a wide variety of gases, and noted the particular importance of carbon dioxide and "aqueous vapour"—water—changes in the concentrations of which, he claimed, "may in fact have produced all the mutations of climate which the researches of geologists reveal." In 2000, the University of East Anglia honored this discovery by naming its new climate research facility the Tyndall Centre.

The story is both true and compelling, and it works well as part of a lineage that also includes Jean-Baptiste Joseph Fourier, the French polymath best known for his mathematical work but also for his recognition in the 1820s that the atmosphere works to redistribute and maintain the temperature of the earth, and Svante Arrhenius, the Swedish Nobel Laureate and then recent divorcé whose mathematical calculations of CO_2 doubling set a standard against which climate models would be compared for the next century.[3]

But there is much more going on in "On the Absorption and Radiation of Heat by Gases and Vapours" than this rather simple lineage would suggest, and much of it has to do with a part of the history of climate change that a narrow focus on the greenhouse effect does more to mask than it does to reveal. In particular, Tyndall's 1861 paper represents an important salvo in an ongoing battle among European scientists over the nature of the laws of thermodynamics—laws without which none of the twentieth- or twenty-first-century debate over CO_2-induced climate change would make much sense. In both his 1861 paper and an 1862 follow-up, Tyndall's comments on climate are sparse and speculative compared to his discussion of the implications of his experiments for the "dynamical theory of heat," a theory articulated by Tyndall's "old scientific antagonist," William Thomson, which included the conservation of energy (the first law of thermodynamics) and an irreversible "law of dissipation," or entropy, that Tyndall feared reflected the Presbyterian sentiments of some of its acolytes. Tyndall, a self-described "scientific naturalist" who believed in a mechanistic, reducible universe explicable by the atomic theory of matter and his friend Charles Darwin's theory of evolution, saw in his experiments on gaseous absorption a way to investigate the relationship between energy and atoms. To put it simply, what Tyndall cared about here was heat.

Tyndall's interest in heat is not surprising if we understand him as a nineteenth-century physicist. The nature of heat was perhaps the most important question in nineteenth-century physics. In fact, Fourier, too, came to his observations about the action of the atmosphere on terrestrial temperatures in a work that was primarily an investigation of the nature of heat. Fourier hoped to be "the Newton of heat," and he described his commitment to discovering the mathematical laws of heat as a way of understanding the *geophysical* properties of the earth—a novel idea and a term that would not appear in English until nearly the end of the nineteenth century, when not just physicists but also geologists and chemists like Svante Arrhenius would attempt to apply the laws of their disciplines to the large-scale, long-term phenomena of the entire earth. Thus, while from a twenty-first-century perspective Fourier, Tyndall, and Arrhenius seem important because of their specific contributions to the intellectual biography of global warming, approaching their scientific work with nineteenth-century questions in mind gives us insight into questions about a whole new way of thinking about global processes—questions about geophysics—certainly as meaningful to twentieth- and twenty-first-century climatic concerns as the absorption spectra for which the Tyndall Centre honors its namesake.

If too narrow a focus on the lineage of scientific concepts can lead us astray, how should we approach documents from science's past that are often famous for their present-day relevance? That is: how does a historian read a scientific paper as a historical document?

One key to understanding scientific papers of the past on their own terms is to think about how scientists, then and now, read papers. While a historian might read a diary entry, a letter, or a piece of literature straight through from beginning to end, scientists often read scientific papers backwards, beginning not with introductory text but with figures and data, the abstract, and the paper's conclusion. Not all scientists read and write papers this way, but tables, graphs, and visualizations of data often represent the most important portion of a scientific work and reflect the lion's share of the scientists' labor, and starting with figures can give a sense of what is important in a scientific paper.

For example, Tyndall put a great deal of thought and work into the instrumentation he used to measure gases in his laboratory. What can the plate at the end of the paper and the explanation of the ratio photospectrometer tell us about the types of questions he hoped to ask and the means by which he asked them? (Can you figure out how the instrument actually worked?) Like many twenty-first-century scientists, Tyndall may have expected his scientific readers to focus on his figures and his concluding section, engaging with the body of the text only to find more information about experimental design and subordinate points of interest. What points of interest come out in the text, and how? For example, compare Tyndall's short line about CO_2 and climate change (buried on page 28 of the paper) with his conclusion (excerpted from the three full pages of commentary), "On the Physical Connexion between Radiation, Absorption, and Conduction." How are the two parts of the paper different, and what purposes do they serve for the author? Compare these, in turn, with the way Svante Arrhenius broaches the subject of the influence of "carbonic acid in the air" on temperatures on the ground, and the way G. S. Callendar introduces climate change in his much more familiar 1938 paper, "The Artificial Production of Carbon Dioxide and Its Influence on Temperature." Beyond the citations, what subjects and ideas recognizably carry from one to the next, and what changes?

For modern historians, taking historical scientists on their own terms requires that we put the questions that pique our interest—in this case, Tyndall's discussion of CO_2 and water vapor—in conversation with the primary interests of scientists themselves. To mistake their historical concerns for our own is to miss the extent to which the nineteenth-century study of climate change provided not just a specific set of experiments illuminating the greenhouse effect, but rather a whole system of knowledge about how the world works that was—and still is—essential to making the greenhouse effect meaningful.

NOTES

1 Gale E. Christianson, *Greenhouse: The 200-Year Story of Global Warming* (New York: Walker, 1999); James Rodger Fleming, *Historical Perspectives on Climate Change* (New York: Oxford University Press, 1998); Spencer R. Weart, *The*

  Discovery of Global Warming, rev. ed. (Cambridge, MA: Harvard University Press, 2008).

2 For a more detailed analysis of the problem of taking nineteenth-century climate science on its own terms, see Joshua P. Howe, "Getting Past the Greenhouse: John Tyndall and the Nineteenth-Century History of Climate Change," in *The Age of Scientific Naturalism: John Tyndall and His Contemporaries*, ed. Bernard Lightman and Michael Reidy (London: Pickering and Chatto, 2014).

3 See James R. Fleming, "Joseph Fourier, the 'Greenhouse Effect,' and the Quest for a Universal Theory of Terrestrial Temperatures," *Endeavour* 23, no. 2 (1999); James R. Fleming, "Joseph Fourier's Theory of Terrestrial Temperatures," in *Historical Perspectives*, 55–64.

JOSEPH FOURIER

GENERAL REMARKS ON THE TEMPERATURES OF THE GLOBE AND THE PLANETARY SPACES

One of the more colorful characters in the history of nineteenth-century science, Jean-Baptiste Joseph Fourier is often credited with the first usage of the greenhouse analogy in describing the way the atmosphere absorbs and redistributes heat. Fourier was many things—at various points he worked as mathematician at the famous École Polytechnique, served as prefect of Isère and Rhône, acted as special adviser to Napoleon in Egypt, and spent time as a political prisoner of both the revolutionary and reactionary governments of the French Revolution (both times for being a lukewarm revolutionary). He was not, however, a prophet of climate change, and in fact Fourier was only marginally concerned with what we now call the greenhouse effect. His discussion of the earth's climate fit into a larger project to articulate the action of heat in mathematical terms. His description here of Henri de Saussure's experiments with the heliothermometer—which he likened in French to *une serre*, or a greenhouse—reveal a nascent understanding of the way the earth absorbs and reradiates heat at different wavelengths, which is in fact quite important to understanding the earth's energy budget. (Fourier's "non-luminous radiating heat" is familiar to modern scientists as infrared radiation). But the phenomenon he hoped to explore was different from the mechanism by which the atmosphere warms the globe, and the specific "greenhouse" reference may thus be a red herring that distracts us from a much larger contribution to a broader scientific development,

---

Joseph Fourier, "General Remarks on the Temperatures of the Globe and the Planetary Spaces," trans. Ebenezer Burgess, *American Journal of Science* 32 (1837): 1–20. Excerpt: pp. 1–2, 10–12, 13, 19–20.

the advent of what he called geophysics, which opened the door to the science of climate change in the twentieth century.*

• •

The question of terrestrial temperature, one of the most remarkable and difficult in natural philosophy, involves very different elements which require to be considered in a general light. I have thought it would be useful to have condensed in a single essay, all the results of this theory. The analytical details here admitted, are found in works which I have already published. I was specially desirous of presenting to philosophers, in a concise table, a complete view of the phenomena and the mathematical relations which exist between them.

The heat of the earth is derived from three sources, which should first be distinctly mentioned.

1. The earth is heated by the solar rays; the unequal distribution of which causes diversities of climate.
2. It partakes of the common temperature of the planetary spaces; being exposed to the radiations from the innumerable stars which surround the solar system.
3. The earth preserves in its interior a part of that primitive heat which it had at the time of the first formation of the planets.

We shall separately examine each of these three causes, and the phenomena which they produce. We will show, as clearly as we are able in the present state of the science, the principal features of these phenomena. For the purpose of giving a general idea of this great question, and showing at a glance the results of our researches, we present them

---

* Fourier's greenhouse analogy referred to the way the heliothermometer prevented warm air from escaping into the atmosphere, which is, in part, how a greenhouse works. He did not suggest that the atmosphere works in this way, however. In fact, the term *greenhouse effect* is a curious misnomer. Rather than trapping warm air close to the earth, the main mechanism of the climatic greenhouse effect occurs as the atmosphere itself—the glass of the greenhouse, if one pursues the analogy—actually absorbs and reradiates heat, which is not at all how an actual greenhouse works.

in the following summary, which is in some measure a synoptic table of the contents of this article, and of several which have preceded it.

The solar system is situated in a region of the universe, every point of which has a common and constant temperature, determined by the rays of light and heat which proceed from the surrounding stars. This low temperature of the planetary space, is a little below that of the polar regions of the earth. The earth would have only the same temperature with the heavens, were it not for two causes which are concurring to heat it. One is the internal heat which it possessed at its formation, a part of which only is dissipated through the surface; the other is the continued action of the solar rays, which penetrate the whole mass, and produce at the surface, the diversities of climate. . . .

We can determine with some degree of precision, the temperature which the earth would have acquired if situated in the place of each of the planets; but the temperature of the planets themselves, cannot be ascertained; for in order to that we must know the state of the surface and the atmosphere. . . . The motion of the air and waters, the extent of the seas, the elevation and form of the surface, the effects of human industry and all the accidental changes of the earth's surface, modify the temperatures of each climate. . . .

The motion of the waters and of the air, tends to modify the effects of heat and cold.

It renders their distribution more uniform, but it would be impossible for the atmosphere to supply the place of that universal cause which supports the common temperature of the planetary spaces; and if this cause did not exist, we should observe, notwithstanding the atmosphere and seas, an enormous difference between the temperatures of the equatorial and polar regions.

It is difficult to know how far the atmosphere influences the mean temperature of the globe; and in this examination we are no longer guided by a regular mathematical theory. It is to the celebrated traveller, M. de Saussure, that we are indebted for a capital experiment, which appears to throw some light on this question.

The experiment consists in exposing to the rays of the sun, a vessel covered with one or more plates of glass, very transparent, and placed at some distance one above the other. The interior of the vessel is furnished with a thick covering of black cork, proper for receiving and preserving heat. The heated air is contained in all parts, both in the interior of the vessel and in the spaces between the plates. Thermometers placed in the vessel itself and in the intervals above, mark the degree of heat in each space. This instrument was placed in the sun about noon, and the thermometer in the vessel was seen to rise to 70°, 80°, 100°, 110°, (Reaumur,) and upwards. The thermometers placed in the intervals between the glass plates indicated much lower degrees of heat, and the heat decreased from the bottom of the vessel to the highest interval.

The effect of solar heat upon air confined within transparent coverings, has long since been observed. The object of the apparatus we have just described, is to carry the acquired heat to its maximum; and especially to compare the effect of the solar ray upon very high mountains, with what is observed in plains below. This experiment is chiefly worthy of remark on account of the just and extensive inferences drawn from it by the inventor. It has been repeated several times at Paris and Edinburgh, and with analogous results.

The theory of the instrument is easily understood. It is sufficient to remark, 1st, that the acquired heat is concentrated, because it is not dissipated immediately by renewing the air; 2d, that the heat of the sun, has properties different from those of heat without light. The rays of that body are transmitted in considerable quantity through the glass plates into all the intervals, even to the bottom of the vessel. They heat the air and the partitions which contain it. Their heat thus communicated ceases to be luminous, and preserves only the properties of non-luminous radiating heat. In this state it cannot pass through the plates of glass covering the vessel.

It is accumulated more and more in the internal which is surrounded by substances of small conducting power, and the temperature rises till the heat flowing in, shall exactly equal that which is dissipated. . . .

The mobility of the air, which is rapidly displaced in every direction, and which rises when heated, and the radiation of non-luminous heat into the air, diminish the intensity of the effects which would take place in a transparent and solid atmosphere, but do not entirely change their character. The decrease of heat in the higher regions of the air does not cease, and the temperature can be augmented by the interposition of the atmosphere, because heat in the state of light finds less resistance in penetrating the air, than in repassing into the air when converted into non-luminous heat. We shall now consider that peculiar heat which our globe had at the time of the formation of the planets, and which continues to be dissipated at the surface under the influence of the low temperature of the planetary space. . . .

I have united in this article all the principal elements of the analysis of terrestrial temperature. It is made up from the results of my researches long since given to the public. . . . The object of this last article is to invite attention to one of the most important questions of natural philosophy, and to present general views and results. It would be impossible to resolve all doubts connected with a subject so extensive; which comprises, besides the results of a new and different analysis, physical considerations very varied in their natures. Exact observations will hereafter be multiplied. The laws on which depends the motion of heat in liquids and air, will be studied. Perhaps other properties of radiating heat will be discovered, or causes which modify the temperatures of the globe. But all the principal laws of the motion of heat are known. This theory, which rests upon immutable foundations, constitutes a new branch of mathematical science. It is composed, at present, of differential equations of the motion of heat in solids and liquids, and of the integrals of these first equations, and theorems relative to the equilibrium of radiating heat.

JOHN TYNDALL

THE BAKERIAN LECTURE

On the Absorption and Radiation of Heat by Gases and Vapours, and on the Physical Connexion of Radiation, Absorption, and Conduction

Tyndall presented his 1859–1861 research on gaseous absorption (based on what we would now call opacity) before the Royal Society in 1861 as the Bakerian Lecture, the society's prestigious annual prize lecture in the physical sciences. His conclusions supported a version of the atomic theory of matter espoused by his Royal Society colleagues, a theory that stood alongside the theory of evolution by natural selection and the reversibility of energy processes at the heart of a secularist worldview called scientific naturalism. More than anything, however, the paper stood out for its methodology—and particularly for its chief instrument of investigation, the ratio photospectrometer. What does the plate at the end of the paper tell us about Tyndall's interests? And how should we interpret his brief sojourn into the implications of carbonic acid (carbon dioxide for a modern reader) and water vapor in the atmosphere next to his discussion of the mechanisms of heat?*

• •

John Tyndall, "The Bakerian Lecture: On the Absorption and Radiation of Heat by Gases and Vapours, and on the Physical Connexion of Radiation, Absorption, and Conduction," *Philosophical Magazine* 4, no. 22 (1861): 169–94, 273–85. Excerpt: pp. 1, 6–7, 28–29, 33–34, 35, plate 1.

* For more on scientific naturalism, see Bernard V. Lightman and Reidy, *The Age of Scientific Naturalism: Tyndall and His Contemporaries*, 2014; Bernard Lightman and Gowan Dawson, *Victorian Scientific Naturalism: Community, Identity,*

The researches on glaciers which I have had the honour of submitting from time to time to the notice of the Royal Society, directed my attention in a special manner to the observations and speculations of De Saussure, Fourier, M. Pouillet, and Mr. Hopkins, on the transmission of solar and terrestrial heat through the earth's atmosphere. This gave practical effect to a desire which I had previously entertained to make the mutual action of radiant heat and gases of all kinds the subject of an experimental inquiry.

Our acquaintance with this department of Physics is exceedingly limited. . . .

§4.

The entire apparatus made use of in the experiments on absorption is figured on Plate I. S S′ is the *experimental tube* composed of brass, polished within, and connected, as shown in the figure, with the air-pump, A A. At S and S′ are the plates of rock-salt which close the tube air-tight. The length from S to S′ is 4 feet. C is a cube containing boiling water, in which is immersed the thermometer t. The cube is of cast copper, and on one of its faces a projecting ring was cast to which a brass tube of the same diameter as S S′ and capable of being connected air-tight with the latter, was carefully soldered. The face of the cube within the ring is the radiating plate, which is coated with lampblack. Thus between the cube C and the first plate of rock-salt there is a front chamber F, connected with the air-pump by the flexible tube D D, and capable of being exhausted independently of S S′. To prevent the heat of conduction from reaching the plate of rock-salt S, the tube F is caused to pass through a vessel V, being soldered to the latter where it

---

Continuity (Chicago: University of Chicago Press, 2014); Bernard Lightman, "Victorian Sciences and Religions: Discordant Harmonies," *Osiris*, 2nd ser., 16 (January 1, 2001): 343–66; On energy and thermodynamics in the nineteenth century, see Crosbie Smith, *The Science of Energy: A Cultural History of Energy Physics in Victorian Britain* (Chicago: University of Chicago Press, 1999). The history of the laws of thermodynamics also sustains one of the most famous essays in the history of nineteenth-century physics. Thomas S. Kuhn, "Energy Conservation as an Example of Simultaneous Discovery," in *Critical Problems in the History of Science* (Madison: University of Wisconsin Press, 1959), 321–56.

enters it and issues from it. This vessel is supplied with a continuous flow of cold water through the influx tube $i\,i$, which dips to the bottom of the vessel; the water escapes through the efflux tube $e\,e$, and the continued circulation of the cold liquid completely intercepts the heat that would otherwise reach the plate S. The cube C is heated by the gas-lamp L. P is the thermo-electric pile placed on its stand at the end of the experimental tube, and furnished with two conical reflectors, as shown in the figure. C' is the *compensating cube*, used to neutralize by its radiation the effect of the rays passing through S S'. The regulation of this neutralization was an operation of some delicacy; to effect it the double screen H was connected with a winch and screw arrangement, by which it could be advanced or withdrawn through extremely minute spaces. For this most useful adjunct I am indebted to the kindness of my friend Mr. Gassiot. N N is the galvanometer with perfectly astatic needles, and perfectly non-magnetic coil; it is connected with the pile P by the wires $w\,w$; Y Y is a system of six chloride-of-calcium tubes, each 32 inches long; R is a U-tube containing fragments of pumice-stone, moistened with strong caustic potash; and Z is a second similar tube, containing fragments of pumice-stone wetted with strong sulphuric acid. When drying only was aimed at, the pot as tube was suppressed. When, on the contrary, as in the case of atmospheric air, both moisture and carbonic acid were to be removed, the potash tube was included. G G is a holder from which the gas to be experimented with was sent through the drying tubes, and thence through the pipe $p\,p$ into the experimental tube S S'. The appendage at M and the arrangement at O O may for the present be disregarded; I shall refer to them particularly by and by.

The mode of proceeding was as follows:—The tube S S' and the chamber F being exhausted as perfectly as possible, the connexion between them was intercepted by shutting off the cocks m, m'. The rays from the interior blackened surface of the cube C passed first across the vacuum F, then through the plate of rock-salt S, traversed the experimental tube, crossed the second plate S', and being concentrated by the anterior conical reflector, impinged upon the adjacent face of the pile P. Meanwhile the rays from the hot cube C' fell upon the opposite face of the pile, and the position of the galvanometer needle declared at once which source was predominant. A movement of the screen H back or forward with the hand sufficed to establish an approximate quality; but to make the radiations perfectly equal, and thus bring the needle exactly to o, the fine motion of the screw above referred to was necessary. The needle being at 0° the gas to be examined was admitted into the tube; passing, in the first place, through the drying apparatus. Any required quantity of the gas may be admitted; and here experiments on gases and vapours enjoy an advantage over those with liquids and solids, namely, the capability of changing the density at pleasure. When the required quantity of gas had been admitted, the galvanometer was observed, and from the deflection of its needle the absorption was accurately determined.

Up to about its 36th degree, the degrees of my galvanometer are all equal in value; that is to say, it requires the same amount of heat to move the needle from 1° to 2° as to move it from 35° to 36°. Beyond this limit the degrees are equivalent to larger amounts of heat. The instrument was accurately calibrated by the method recommended by Melloni (Thermochrose, p. 59), so that the precise value of its larger deflections are at once obtained by reference to a table. Up to the 36th degree, therefore, the simple deflections may be regarded as the expression of the absorption, but beyond this the absorption equivalent to ally deflection is obtained from the table of calibration. . . .

§8.

I have now to refer briefly to a point of considerable interest as regards the effect of our atmosphere on solar and terrestrial heat. In examining the separate effects of the air, carbonic acid, and aqueous vapour of the atmosphere on the 20th of last November, the following results were obtained:

Air sent through the system of drying tubes and through the caustic potash tube produced an absorption of about

1.

Air direct from the laboratory, containing therefore its carbonic acid and aqueous vapour, produced an absorption of

15.

Deducting the effect of the gaseous acids, it was found that the quantity of aqueous vapour diffused through the atmosphere on the day in question, produced an absorption at least equal to thirteen times that of the atmosphere itself.

It is my intention to repeat and extend these experiments on a future occasion; but even at present conclusions of great importance may be drawn from them. It is exceedingly probable that the absorption of the solar rays by the atmosphere, as established by M. Pouillet, is mainly due to the watery vapour contained in the air. The vast difference between the temperature of the sun at midday and in the evening, is also probably due in the main to that comparatively shallow stratum of aqueous vapour which lies close to the earth. At noon the depth of it pierced by the sunbeams is very small; in the evening very great in comparison.

The intense heat of the sun's direct rays on high mountains is not, I believe, due to his beams having to penetrate only a small depth of air, but to the comparative absence of aqueous vapour at those great elevations.

But this aqueous vapour, which exercises such a destructive action on the obscure rays, is comparatively transparent to the rays of light. Hence the differential action, as regards the heat coming from the sun to the earth, and that radiated from the earth into space, is vastly augmented by the aqueous vapour of the atmosphere. De Saussure, Fourier, M. Pouillet, and Mr. Hopkins regard this interception of the terrestrial rays as exercising the most important influence on climate. Now if, as the above experiments indicate, the chief influence be exercised by the aqueous vapour, every variation of this constituent must produce a change of climate. Similar remarks would apply to the carbonic acid diffused through the air; while an almost inappreciable admixture of any of the hydrocarbon vapours would produce great effects on the terrestrial rays and produce corresponding changes of climate [Tyndall's *hydrocarbon vapours* refers to methane and related gases]. It is not therefore necessary to assume alterations in the density and height of the atmosphere, to account for different amounts of heat being preserved to the earth at different times; a slight change in its variable constituents would suffice for this. Such changes in fact may have produced all the mutations of climate which the researches of geologists reveal. However this may be, the facts above cited remain; they constitute true causes, the extent alone of the operation remaining doubtful. . . .

§10. ON THE PHYSICAL CONNEXION OF RADIATION, ABSORPTION, AND CONDUCTION

Notwithstanding the great accessions of late years to our knowledge of the nature of heat, we are as yet, I believe, quite ignorant of the atomic conditions on which radiation, absorption, and conduction depend. What are the specific qualities which cause one body to radiate copiously and another feebly? Why, on theoretic grounds, must the equivalence of radiation and absorption exist? Why should a highly diathermanous body, as shown by Mr. Balfour Stewart, be a bad radiator, and an adiathermanous body a good radiator? How is heat conducted? and what is the strict physical meaning of good conduction and bad conduction? Why should good conductors be, in general, bad radiators, and bad conductors good radiators? These, and other questions, referring to facts more or less established, have still to receive their complete answers. It is less with a hope of furnishing such than of shadowing forth the possibility of uniting these various effects by a common bond, that I submit the following reflections to the notice of the Royal Society.

In the experiments recorded in the foregoing pages, we have dealt with free atoms, both simple and compound, and it has been found that in all cases absorption takes place. The meaning of this, according to the dynamical theory of heat, is that no atom is capable of existing in vibrating ether without accepting a portion of its motion. We may, if we wish, imagine a certain roughness of the surface of the atoms which enables the ether to bite them and carry the atom along with it. But no matter what the quality may be which enables any atom to accept motion from the agitated ether, the same quality must enable it to impart motion to still ether when it is plunged in the latter and agitated. It is only necessary to imagine the case of a body immersed in water to see that this must be the case. There is a polarity here as rigid as that of magnetism. From the existence of absorption, we may on theoretic grounds infallibly infer a capacity for radiation; from the existence of radiation, we may with equal certainty infer a capacity for absorption; and each of them must be regarded as the measure of the other.

This reasoning, founded simply on the mechanical relations of the ether and the atoms immersed in it, is completely verified by experiment.

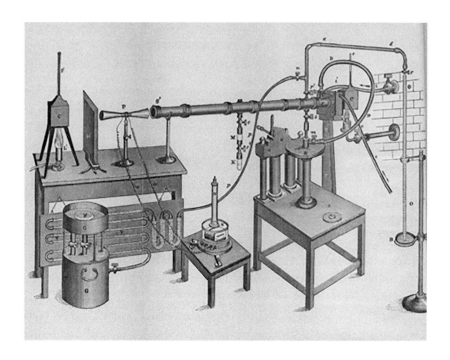

Great differences have been shown to exist among gases as to their powers of absorption, and precisely similar differences as regards their powers of radiation. But what specific property is it which makes one free molecule a strong absorber, while another offers scarcely any impediment to the passage of radiant heat? I think the experiments throw some light upon this question. . . .

Many chemists, I believe, are disposed to reject the idea of an atom, and to adhere to that of equivalent proportions merely. They figure the act of combination as a kind of interpenetration of one substance by another. But this is a mere masking of the fundamental phenomenon. The value of the atomic theory consists in its furnishing the physical explanation of the law of equivalents;—assuming the one the other follows; and assuming the act of chemical union as Dalton figured it, we see that it blends harmoniously with the perfectly independent conception of an ether, and enables us to reduce the phenomena of radiation and absorption to the simplest mechanical principles.

SVANTE ARRHENIUS

ON THE INFLUENCE OF CARBONIC ACID IN THE AIR UPON THE TEMPERATURE OF THE GROUND

In 1896, Swedish Nobel Laureate chemist Svante Arrhenius revisited Tyndall's idea that the absorption of energy by water vapor and CO_2 might help regulate the temperature of the earth. Depressed after an acrimonious divorce and an unsuccessful bid for custody of his son, Arrhenius set to work on a tedious set of mathematical calculations based on the absorption coefficients of gases to create a pencil-and-paper model of the influence of atmospheric CO_2 on the temperature of the earth at various latitudes. Arrhenius was most concerned with using the model to suggest that reduced atmospheric CO_2 accompanied—and potentially caused—the then recently discovered ice age of the geologic past. But his tables and discussion here also reveal temperature predictions for a world of rising CO_2—predictions that have earned him a place among the so-called forefathers of climate science. How does Arrhenius interpret Tyndall in this paper? What is familiar and what is strange about the way he deals with "carbonic acid"?*

Svante Arrhenius, "On the Influence of Carbonic Acid in the Air upon the Temperature of the Ground," *Philosophical Magazine and Journal of Science* 5, no. 41 (April 1896): 237–76. Excerpt: pp. 237–38, 239, 263–69.

* For more on Arrhenius, see James R. Fleming, "John Tyndall, Svante Arrhenius, and Early Research on Carbon Dioxide and Climate," in *Historical Perspectives on Climate Change* (New York: Oxford University Press, 1998), 8:65–68. Arrhenius expanded upon the potential benefits of CO_2-induced warming in a later popular work, *Worlds in the Making: The Evolution of the Universe* (New York: Harper, 1908).

I. INTRODUCTION: OBSERVATIONS OF LANGLEY ON ATMOSPHERICAL ABSORPTION

A great deal has been written on the influence of the absorption of the atmosphere upon the climate. Tyndall in particular has pointed out the enormous importance of this question. To him it was chiefly the diurnal and annual variations of the temperature that were lessened by this circumstance. Another side of the question, that long has attracted the attention of physicists, is this: Is the mean temperature of the ground in any way influenced by the presence of heat-absorbing gases in the atmosphere? Fourier maintained that the atmosphere acts like the glass of a hothouse, because it lets through the light rays of the sun but retains the dark rays from the ground. This idea was elaborated by Pouillet; and Langley was by some of his researches led to the view, that "the temperature of the earth under direct sunshine, even though our atmosphere were present as now, would probably fall to $-200°$ C., if that atmosphere did not possess the quality of selective absorption." . . .

. . . Tyndall held the opinion that the water-vapour has the greatest influence, whilst other authors, for instance Lecher and Penter, are inclined to think that the carbonic acid plays the more important part. The researches of Paschen show that these gases are both very effective, so that probably sometimes the one, sometimes the other, may have the greater effect according to the circumstances. . . .

IV. CALCULATION OF THE VARIATION OF TEMPERATURE THAT WOULD ENSUE IN THE CONSEQUENCES OF A GIVEN VARIATION OF THE CARBONIC ACID IN THE AIR

We now possess all the necessary data for an estimation of the effect on the earth's temperature which would be the result of a given variation of the aerial carbonic acid. . . .

In order to obtain values for the temperature for the whole earth, I have calculated from Dr. Buchan's charts of the mean temperature at different places in every month the mean temperature in every district

TABLE VI.—*Mean Temperature, Relative and Absolute Humidity**.

Latitude	Mean Temperature					Mean Relative Humidity					Mean Absolute Humidity				
	Dec.–Feb.	March–May	June–Aug.	Sept.–Nov.	Mean of the year.	Dec.–Feb.	March–May	June–Aug.	Sept.–Nov.	Mean of the year.	Dec.–Feb.	March–May	June–Aug.	Sept.–Nov.	Mean of the year.
70	−21·1	− 8·3	+ 7·5	− 6·0	− 7·0	86	81	77	84	82	1·15	2·14	6·22	2·84	3·09
60	−11·2	+ 0·2	+13·5	+ 2·2	+ 1·2	83	74	76	80	78·2	2·22	3·82	8·82	4·7	4·9
50	− 1·4	+ 7·8	+18·7	+ 9·7	+ 8·7	78	73	69	76	74	3·86	5·98	10·8	7·16	6·95
40	+ 8·4	+14·5	+21·8	+16·6	+15·3	73	68	67	71	69·7	6·53	8·63	13·4	10·13	9·7
30	+17·0	+21·5	+26·0	+23·0	+21·9	71	68	70	73	70·5	10·36	12·63	17·1	15·0	13·8
20	+23·2	+25·5	+26·8	+25·9	+25·4	74	73	78	77	75·5	15·3	17·0	19·6	16·8	17·2
10	+25·5	+25·8	+25·4	+25·5	+25·5	77	78	82	81	79·5	17·7	18·9	19·9	19·3	18·9
0	+25·7	+25·5	+24·0	+25·0	+25·1	81	81	82	80	81	19·4	19·0	17·9	18·3	18·7
−10	+24·9	+24·0	+20·8	+23·1	+23·2	79	78	80	77	78·5	18·0	17·1	14·6	16·0	16·4
−20	+22·4	+20·5	+16·4	+19·3	+19·7	75	79	80	75	77·2	14·8	14·0	11·1	13·0	13·2
−30	+17·5	+15·2	+11·3	+14·2	+14·5	75	80	80	79	78·5	11·1	10·4	8·1	9·6	9·8
−40	+11·6	+ 9·5	+ 5·9	+ 8·2	+ 8·7	81	81	83	79	81	8·34	7·08	5·94	6·63	6·99
−50	+ 5·3	+ 2·0	− 0·4	+ 1·6	+ 2·1	83	79	—	—	—	5·74	4·46	—	—	—

* From the figures for temperature and relative humidity I have calculated the absolute humidity in grams per cubic metre.

that is contained between two parallels differing by 10 and two meridians, differing by 20 degrees (*e.g.* between 0° and 10° N. and 160° and 180° W).

By means of these values, I have calculated the mean alteration of the temperature that would follow if the quantity of carbonic acid varied from its present mean value (K = 1) to another, viz. to K = 0.671, 1.5, 2, 2.5, and 3 respectively. This calculation is made for every tenth parallel, and separately for the four seasons of the year. The variation is given in Table VII.

A glance at this Table shows that the influence is nearly the same over the whole earth. The influence has a minimum near the equator, and increases from this to a flat maximum that lies the further from the equator the higher the quantity of carbonic acid in the air. For K = 0.67 the maximum effect lies about the 40th parallel, for K = 1.5 on the 50th, for K = 2 on the 60th, and the higher K-values above the 70th parallel. The influence is in general greater in the winter than in the summer, except in the case of the parts that lie between the maximum and the

TABLE VII.—*Variation of Temperature caused by a given Variation of Carbonic Acid.*

Latitude	Carbonic Acid=0.67					Carbonic Acid=1.5					Carbonic Acid=2.0					Carbonic Acid=2.5					Carbonic Acid=3.0				
	Dec.–Feb.	March–May	June–Aug.	Sept.–Nov.	Mean of the year	Dec.–Feb.	March–May	June–Aug.	Sept.–Nov.	Mean of the year	Dec.–Feb.	March–May	June–Aug.	Sept.–Nov.	Mean of the year	Dec.–Feb.	March–May	June–Aug.	Sept.–Nov.	Mean of the year	Dec.–Feb.	March–May	June–Aug.	Sept.–Nov.	Mean of the year
70	-2.9	-3.0	-3.4	-3.1	-3.1	3.3	3.4	3.8	3.6	3.52	6.0	6.1	6.0	6.1	6.05	7.9	8.0	7.9	8.0	7.95	9.1	9.3	9.4	9.4	9.3
60	-3.0	-3.2	-3.4	-3.3	-3.22	3.4	3.7	3.6	3.8	3.62	6.1	6.1	5.8	6.1	6.02	8.0	8.0	7.6	7.9	7.87	9.3	9.5	8.9	9.5	9.3
50	-3.2	-3.3	-3.3	-3.4	-3.3	3.7	3.8	3.4	3.7	3.65	6.1	6.1	5.5	6.0	5.92	8.0	7.9	7.0	7.9	7.7	9.5	9.4	8.6	9.2	9.17
40	-3.4	-3.4	-3.2	-3.3	-3.32	3.7	3.6	3.3	3.5	3.52	6.0	5.8	5.4	5.6	5.7	7.9	7.6	6.9	7.3	7.42	9.3	9.0	8.2	8.8	8.82
30	-3.3	-3.2	-3.1	-3.1	-3.17	3.5	3.3	3.2	3.5	3.47	5.6	5.4	5.0	5.2	5.3	7.2	7.0	6.6	6.7	6.87	8.7	8.3	7.5	7.9	8.1
20	-3.1	-3.1	-3.0	-3.1	-3.07	3.5	3.2	3.1	3.2	3.25	5.2	5.0	4.9	5.0	5.02	6.7	6.6	6.3	6.6	6.52	7.9	7.5	7.2	7.5	7.52
10	-3.1	-3.0	-3.0	-3.0	-3.02	3.2	3.2	3.1	3.1	3.15	5.0	5.0	4.9	4.9	4.95	6.6	6.4	6.3	6.4	6.42	7.4	7.3	7.2	7.3	7.3
0	-3.0	-3.0	-3.1	-3.0	-3.02	3.1	3.1	3.2	3.2	3.15	4.9	4.9	5.0	5.0	4.95	6.4	6.4	6.6	6.6	6.5	7.3	7.3	7.4	7.4	7.35
-10	-3.1	-3.1	-3.2	-3.1	-3.12	3.2	3.2	3.2	3.2	3.2	5.0	5.0	5.2	5.1	5.07	6.6	6.6	6.7	6.7	6.65	7.4	7.5	8.0	7.6	7.62
-20	-3.1	-3.2	-3.3	-3.2	-3.2	3.2	3.2	3.4	3.3	3.27	5.2	5.3	5.5	5.4	5.35	6.7	6.8	7.0	7.0	6.87	7.9	8.1	8.6	8.3	8.22
-30	-3.3	-3.3	-3.4	-3.4	-3.35	3.4	3.5	3.7	3.5	3.52	5.5	5.6	5.8	5.6	5.62	7.0	7.2	7.7	7.4	7.32	8.6	8.7	9.1	8.8	8.8
-40	-3.4	-3.4	-3.3	-3.4	-3.37	3.6	3.7	3.8	3.7	3.7	5.8	6.0	6.0	6.0	5.95	7.7	7.9	7.9	7.9	7.85	9.1	9.2	9.4	9.3	9.25
-50	-3.2	-3.3	—	—	—	3.8	3.7	—	—	—	6.0	6.1	—	—	—	7.9	8.0	—	—	—	9.4	9.5	—	—	—

pole. The influence will also be greater the higher the value of v, that is in general somewhat greater for land than for ocean. On account of the nebulosity of the Southern hemisphere, the effect will be less there than in the Northern hemisphere. An increase in the quantity of carbonic acid will of course diminish the difference in temperature between day and night. A very important secondary elevation of the effect will be produced in those places that alter their albedo by the extension or regression of the snow-covering (see p. 257), and this secondary effect will probably remove the maximum effect from lower parallels to the neighborhood of the poles. . . .

V. GEOLOGICAL CONSEQUENCES

I should certainly not have undertaken these tedious calculations if an extraordinary interest had not been connected with them. In the Physical Society of Stockholm there have been occasionally very lively discussions on the probable causes of the Ice Age; and these discussions have, in my opinion, led to the conclusion that there exists as yet no satisfactory hypothesis that could explain how the climatic conditions for an ice age could be realized in so short a time as that which has elapsed from the days of the glacial epoch. The common view hitherto has been that the earth has cooled in the lapse of time; and if one did not know that the reverse has been the case one would certainly assert that this cooling must go on continuously. Conversations with my friend and colleague Professor Hogbom, together with the discussions above referred to, led me to make a preliminary estimate of the probable effect of a variation of the atmospheric carbonic acid on the temperature of the earth. As this estimation led to the belief that one might in this way, probably find an explanation for temperature variation of $5°-10°C$. I worked out the calculation in more detail and lay it now before the public and the critics.

From geological researches the fact is well established that in Tertiary times there existed a vegetation and an animal life in the temperate and arctic zones that must have been conditioned by a much higher temperature than the present in the same regions. The temperature in the arctic zones appears to have exceeded the present temperature by about 8 or 9 degrees. To this genial time the ice age succeeded, and this was one or more times interrupted by interglacial periods with a climate of about the same character at the present, sometimes even milder. When the ice age had its greatest extent, the countries that now enjoy the highest civilization were covered with ice. This was the case with Ireland, Britain (except a small part in the south), Holland, Denmark, Sweden and Norway, Russia (to Kiev, Orel, and Nijni Novgorod), Germany and Austria (to the Harz, Erz-Gebirge, Dresden, and Cracow). At the same time an ice-cap from the Alps covered Switzerland, parts of France, Bavaria south of the Danube, the Tyrol, Styria, and other Austrian countries, and descended into the northern part of Italy. Simultaneously, too, North America was covered with ice on the west coast to the 47th parallel, on the east coast to the 50th, and in the central part to the 37th (confluence of the Mississippi and Ohio rivers). In the most different parts of the world, too, we have found traces of a great ice age, as in the Caucasus, Asia Minor, Syria, the Himalayas, India, Thian Shan, Altai, Atlas, on Mount Kenia and Kilimandjaro (both very near to the equator), in South Africa, Australia, New Zealand, Kerguelen, Falkland Islands, Patagonia and other parts of South America. The geologists in general are inclined to think that these glaciations were simultaneous on the whole earth; and this most natural view would probably have been generally accepted if the theory of Croll, which demands a genial age of the Southern hemisphere at the same time as an ice age on the Northern and *vice versa*, had not influenced opinion....

One may now ask, How much must the carbonic acid vary according to our figures, in order that the temperature should attain the same values as in the Tertiary and Ice ages respectively? A simple calculation shows that the temperature in the arctic regions would rise about 8°C to 9°C., if the carbonic acid increased to 2.5 or 3 times its present value. In order to get the temperature of the ice age between the 40th and 50th parallels, the carbonic acid in the air should sink to 0.62–0.55 of its present value (lowering of temperature 4°C–5°C.)....

There is now an important question which should be answered, namely:—Is it probable that such great variations in the quantity of carbonic acid as our theory requires have occurred in relatively short geological time?

G. S. CALLENDAR

THE ARTIFICIAL PRODUCTION OF CARBON DIOXIDE AND ITS INFLUENCE ON TEMPERATURE

A number of scientists articulated CO_2-based theories of climatic change in the late nineteenth and early twentieth centuries, but Guy Stewart Callendar was the first scientist to articulate the basic theory of CO_2-driven anthropogenic warming in its recognizable modern form. The paper was remarkably straightforward, and the mechanism of global warming explained here may look familiar. But how do Callendar's conclusions about its impacts on human life compare to the descriptions provided by Tyndall and Arrhenius? Prescient as it appears, the paper was largely ignored in its own time, for reasons that will become clearer in part 2. How would you compare them to the conclusions of later scientists, either from the next chapter or from today?*

· ·

G. S. Callendar, "The Artificial Production of Carbon Dioxide and Its Influence on Temperature," *Quarterly Journal of the Royal Meteorological Society* 64, no. 275 (April 1938): 223–40. Excerpt: pp. 223–24, 236.

* By far the best work on Callendar comes from James Roger Fleming, *The Callendar Effect: The Life and Times of Guy Stewart Callendar (1898–1964), the Scientist Who Established the Carbon Dioxide Theory of Climate Change* (Boston, MA: American Meteorological Society, 2007).

SUMMARY

By fuel combustion man has added about 150,000 million tons of carbon dioxide to the air during the past half century. The author estimates from the best available data that approximately three quarters of this has remained in the atmosphere.

The radiation absorption coefficients of carbon dioxide and water vapour are used to show the effect of carbon dioxide on "sky radiation." From this the increase in mean temperature, due to the artificial production of carbon dioxide, is estimated to be at the rate of 0.005°C per year at the present time.

The temperature observations at 200 meteorological stations are used to show that world temperatures have actually increased at an average rate of 0.005°C per year during the past half century.

Few of those familiar with the natural heat exchanges of the atmosphere, which go into the making of our climates and weather, would be prepared to admit that the activities of man could have any influence upon phenomena of so vast a scale.

In the following paper I hope to show that such influence is not only possible, but is actually occurring at the present time.

It is well known that the gas carbon dioxide has certain strong absorption bands in the infra-red region of the spectrum, and when this fact was discovered some 70 years ago it soon led to speculation on the effect which changes in the amount of the gas in the air could have on the temperature of the earth's surface. In view of the much larger quantities and absorbing power of atmospheric water vapour it was concluded that the effect of carbon dioxide was probably negligible, although certain experts, notably Svante Arrhenius and T. C. Chamberlin, dissented from this view....

I. THE RATE OF ACCUMULATION OF ATMOSPHERIC CARBON DIOXIDE

I have examined a very accurate set of observations (Brown and Escombe, 1905), taken about the year 1900, on the amount of carbon dioxide in the free air, in relation to the weather maps of the period.

From them I concluded that the amount of carbon dioxide in the free air of the North Atlantic region, at the beginning of this century, was 2.74 ± 0.05 parts in 10,000 by volume of dry air.

A great many factors which influence the carbon cycle in nature have been examined in order to determine the quantitative relation between the natural movements of this gas and the amounts produced by the combustion of fossil fuel. Such factors included the organic deposit of carbon in swamps, etc., the average rate of fixation of the gas by the carbonisation of alkalies from igneous rocks, and so on. The general conclusion from a somewhat lengthy investigation on the natural movements of carbon dioxide was that there is no geological evidence to show that the *net* offtake of the gas is more than a small fraction of the quantity produced from fuel. (The artificial production at present is about 4,500 million tons per year.) . . .

In conclusion it may be said that the combustion of fossil fuel, whether it be peat from the surface or oil from 10,000 feet below, is likely to prove beneficial to mankind in several ways, besides the provision of heat and power. For instance the above mentioned small increases of mean temperature would be important at the northern margin of cultivation, and the growth of favourably situated plants is directly proportional to the carbon dioxide pressure (Brown and Escombe, 1905). In any case the return of the deadly glaciers should be delayed indefinitely.

As regards the reserves of fuel these would be sufficient to give at least ten times as much carbon dioxide as there is in the air at present.

PART 2

THE COLD WAR ROOTS OF GLOBAL WARMING

As we saw in the first part, in 1938, G. S. Callendar read "The Artificial Production of Carbon Dioxide and Its Influence on Temperature" before the Royal Society. Callendar claimed that 1) humans had increased the amount of CO_2 in the atmosphere by burning fossil fuels; 2) increased CO_2 could cause an increase in global temperatures; and 3) global temperatures were, in fact, rising. In retrospect, here was a clear articulation of what we now understand as global warming—perhaps not a dire warning, but at the very least a well-reasoned analysis of humanity's impact on the global energy budget. And yet, even the few scientists familiar with Callendar's work in 1938 did not take it very seriously, and Callendar, like his CO_2 theory of climate, remained relatively obscure for most of his lifetime.

Nineteen years later, an oceanographer named Roger Revelle famously published a paper with radiochemist Hans Suess that effectively confirmed Callendar's thesis that increased CO_2 should lead to an increase in global temperature. Unlike Callendar's paper, however, Revelle's famous declaration that "human beings are now carrying out a large-scale geophysical experiment" via the massive emission of CO_2 in the atmosphere drew attention from around the scientific community and even from politicians and members of the public, and historians have come to understand it as a watershed moment in the modern history of global warming. Why, two decades later, did this paper capture scientific and public attention after Callendar's paper had almost entirely failed to do so?

The common story holds that the Revelle and Suess paper sounded the alarm on global warming—that the combined insight of two diligent scientists awakened a new awareness of the fragility of the global atmosphere. But when you look at the papers side by side, what differences appear between them? How plausible is it that the differences in reception rely solely—or even primarily—on differences in the science? As you might find, in some ways, Revelle and Suess presented a much more cautious picture of CO_2's potential impact on climate than did Callendar, whom the two Scripps Institute of Oceanography scientists cited—and attacked for going *too far* with the links between CO_2 and warming—throughout their paper. In fact, within the scientific community, the meaning of Revelle and Suess's muddled paper itself did not become clear until Bert Bolin and Erik Eriksson revisited its conclusions in a 1959 paper.[1] The difference in response, then, must reflect some other difference in the historical context in which the two papers introduced the question of CO_2-induced warming. Between 1938 and 1957, what had changed?

Revelle and Suess's famous paper does not reveal much about the contextual divide between the amateur meteorologist and the oceanographer-chemist duo. Revelle's contemporaneous appearances before congressional committees on American science, however, provide a window into the conditions that enabled scientists in America to put global warming on the federal map in the 1950s and 1960s. Alongside his scientific work, Revelle's testimony demonstrates the importance of the Cold War in the early political history of global warming. What evidence can you find in these texts for a connection between the Cold War and concerns over climate change?

Congressional hearings present particular advantages and liabilities for scholars of American history, and it is important to understand just what kind of document a hearing produces in order to use testimony like Revelle's as a piece of historical evidence. The floor and meeting rooms of Congress are by their nature forums for political performance. Unlike other authors, witnesses deliver testimony orally to an audience; stenographers turn those oral presentations into written documents after the fact. The record of a hearing is thus not self-consciously a text per se, but rather the transcript of interactions between individuals looking out for particular personal, party, or institutional interests, in conversation with specific people in positions of power. Even hearings about science that primarily feature scientists contain a mixture of basic information; explicit recommendations, requests, complaints, and objections, often themselves a mix of the scientific and political; and a rich array of purely political rhetoric, crafted either explicitly or implicitly to appeal to particular constituencies that serve as sources of power, money, or votes. In short, there is always a lot going on in a congressional hearing, though the subject matter itself may appear relatively mundane.

One way to make sense of the text and subtext of a hearing is to think about it in the way you might think about the text of a play. Often, understanding drama requires that we ask what individual characters want, which is revealed in what they say or do to go about getting it. Start with the basics. Who is in the room (this is often more difficult to figure out than it appears when it comes to members of Congress) and in what capacity? What can you learn about the scene from the hearing's title? What committee, subcommittee, or joint committee hosts the hearing, and how might the legislative context shape or constrain the conversation? What states do the congressional leaders represent on the committee, and how does that shape the way you interpret their remarks? How do witnesses affiliate in this public space, professionally, politically, or institutionally?

Start with Revelle's 1956 hearing. From the title you know that in 1956, Revelle appeared before a congressional appropriations committee dealing with the International Geophysical Year (IGY), an enormous cooperative international scientific research effort that would blossom into a project involving more than sixty thousand scientists and technicians from sixty-six nations.[2] Revelle introduced himself as the director of Scripps Institution of Oceanography, and a historian can safely assume that, perhaps among other objectives, Revelle hoped to use his appearance as a chance to lobby for more federal science funding for oceanography and atmospheric science—part of the larger IGY budget that would eventually find its way to his institution. What other priorities might Revelle have related to his role as Scripps director? Where do you see his administrative role reflected in the text?

Beyond his basic goals, Revelle's rhetorical strategy for achieving that objective also reveals the extent to which the Cold War influenced both the presentation and reception of the CO_2 issue. Clearly, Revelle was actually interested in and concerned about CO_2; he made nearly the same claim in his 1956 testimony about the "great geophysical experiment" that he and Suess would propose the next year in the scientific journal *Tellus*. What differences do you see between the way Revelle presented that "experiment" in a scientific journal and the way he presented it to congressmen like Sidney Yates and Albert Richard Thomas? Both Democrats, Yates and Thomas hoped to make political hay on mounting criticisms of the Republican Eisenhower administration for taking Cold War science too lightly in the late 1950s—criticisms that would, coincidentally, amplify considerably in 1957, when the Soviet Union launched the first artificial satellite, *Sputnik*. Even before *Sputnik*, Eisenhower's most powerful congressional rival, Senator Lyndon Johnson, warned the Democratic caucus that "from space, the masters of infinity could have the power to control the earth's weather, to cause drought and flood, to change the tides and raise the level of the sea, to divert the Gulf Stream, and change the temperate climates to frigid."[3] Speaking in metaphorical terms about scientific discovery, Revelle worried aloud—and for the record—that "the Russian Columbuses have been doing a lot better than we have."[4] How else did Revelle manage to frame uncertainties about CO_2 in geopolitical terms associated with Cold War fears about the perceived superiority of Russian science? To what extent do these concerns also motivate Howard T. Orville's 1958 testimony on weather modification? And how do they show up outside of Congress in official institutional documents like those of the National Science Foundation in its planning for the National Center for Atmospheric Research? More broadly, what can the combination of scientific papers, congressional testimony, and agency documents tell us about the relationship between climate science and the Cold War in the 1950s?

As the 1956 hearings reveal, the Cold War context of Revelle and Suess's findings gave CO_2 and climate change a whole new set of meanings, and as a result, we remember Revelle, not Callendar, as the self-described "granddaddy of global warming."[5] Reading congressional testimony between the lines can help make sense of how and why the meanings of CO_2 changed in the Cold War setting, and how scientists helped shape those changing meanings.

NOTES

1. From a purely scientific perspective, the Bolin and Eriksson paper may be a more important contribution to our understanding of global warming, but the *Tellus* piece, in part because of Revelle's personality and public stature and in part because of the "geophysical experiment" phrasing, has become the more renowned of the two. Bert Bolin and Erik Eriksson, "Distribution of Matter in the Sea and Atmosphere: Changes in Carbon Dioxide Content of the Atmosphere and Ocean Due to Fossil Fuel Combustion," in *The Atmosphere and the Sea in Motion*, ed. Bert Bolin (New York: Rockefeller Institute Press, 1959), 130–42.
2. For more on IGY, see the contemporary report by Walter Sullivan, *Assault on the Unknown: The International Geophysical Year* (New York: McGraw-Hill, 1961); for its origins, see also Fae L. Korsmo, "The Genesis of the International Geophysical Year," *Physics Today* 60, no. 7 (July 2007): 38–43.
3. Howard T. Orville, "The Impact of Weather Control on the Cold War," in *Weather Modification Research*, hearings of House Committee on Interstate and Foreign Commerce, 85th Cong., March 18–19, 1958, 51–53.
4. Roger Revelle, testimony in *National Science Foundation: International Geophysical Year*, hearings of House Committee on Appropriations, 84th Cong., February 6, 1956, 466.
5. For more on the links between the Cold War and global warming, see Spencer Weart, "From the Nuclear Frying Pan into the Global Fire," *Bulletin of the Atomic Scientists* 48, no. 5 (June 1992): 18–27.

ROGER REVELLE AND HANS E. SUESS

CARBON DIOXIDE EXCHANGE BETWEEN ATMOSPHERE AND OCEAN AND THE QUESTION OF AN INCREASE OF ATMOSPHERIC CO_2 DURING THE PAST DECADES

In 1957, Roger Revelle and Hans Suess of the Scripps Institution of Oceanography produced a scientific paper that has become one of the favorite and most important starting points for modern historians of climate change. Published in the oceanographic journal *Tellus*, the body of the Revelle and Suess paper largely focused on a set of interactions between ocean and atmosphere that the authors initially believed would undermine Callendar's 1938 hypothesis on increases in atmospheric CO_2. Just before the paper went to publication, however, the authors began to realize that their equations for ocean-atmosphere interactions left the atmosphere with considerable unaccounted-for CO_2. Revelle and Seuss famously described the contribution of CO_2 to the atmosphere as a "large-scale geophysical experiment," a phrase that would appear over and over again in the making of climate change history.

· ·

Roger Revelle and Hans E. Suess, "Carbon Dioxide Exchange between Atmosphere and Ocean and the Question of an Increase in Atmospheric CO_2 during the Past Decades," *Tellus* 9 (1957): 18–27. Excerpt: pp. 18–20, 24, 26.

ABSTRACT

From a comparison of C^{14}/C^{12} and C^{13}/C^{12} ratios in wood and in marine material and from a slight decrease of the C^{14} concentration in terrestrial plants over the past 50 years it can be concluded that the average lifetime of a CO_2 molecule in the atmosphere before it is dissolved into the sea is of the order of 10 years. This means that most of the CO_2 released by artificial fuel combustion since the beginning of the industrial revolution must have been absorbed by the oceans. The increase of atmospheric CO_2 from this cause is at present small but may become significant during future decades if industrial fuel combustion continues to rise exponentially.

Present data on the total amount of CO_2 in the atmosphere, on the rates and mechanisms of exchange, and on possible fluctuations in terrestrial and marine organic carbon, are inadequate for accurate measurement of future changes in atmospheric CO_2. An opportunity exists during the International Geophysical Year to obtain much of the necessary information.

INTRODUCTION

In the middle of the 19th century appreciable amounts of carbon dioxide began to be added to the atmosphere through the combustion of fossil fuels. The rate of combustion has continually increased so that at the present time the annual increment from this source is nearly 0.4% of the total atmospheric carbon dioxide. By 1960 the amount added during the past century will be more than 15%.

Callendar (1938, 1940, 1949) believed that nearly all the carbon dioxide produced by fossil fuel combustion has remained in the atmosphere, and he suggested that the increase in atmospheric carbon dioxide may account for the observed slight rise of average temperature in northern latitudes during recent decades. He thus revived the hypothesis of T. C. Chamberlin (1899) and S. Arrhenius (1903) that climatic changes may be related to fluctuations in the carbon dioxide content of the air. These authors supposed that an increase of carbon dioxide in the upper atmosphere would lower the mean level of back radiation in the infrared and thereby increase the average temperature near the earth's surface.

Subsequently, other authors have questioned Callendar's conclusions on two grounds. First, comparison of measurements made in the 19th century and in recent years do not demonstrate that there has been a significant increase in atmospheric CO_2 (Slocum, 1955; Fonselius et al., 1956). Most of the excess CO_2 from fuel combustion may have been transferred to the ocean, a possibility suggested by S. Arrhenius (1903). Second, a few percent increase in the CO_2 content of the air, even if it has occurred, might not produce an observable increase in average air temperature near the ground in the face of fluctuations due to other causes. So little is known about the thermodynamics of the atmosphere that it is not certain whether or how a change in infrared back radiation from the upper air would affect the temperature near the surface. Calculations by Plass (1956) indicate that a 10% increase in atmospheric carbon dioxide would increase the average temperature by 0.36°C. But, amplifying or feedback processes may exist such that a slight change in the character of the back radiation might have a more pronounced effect. Possible examples are a decrease in albedo of the earth due to melting of ice caps or a rise in water vapor content of the atmosphere (with accompanying increased infrared absorption near the surface) due to increased evaporation with rising temperature.

During the next few decades the rate of combustion of fossil fuels will continue to increase, if the fuel and power requirements of our worldwide industrial civilization continue to rise exponentially, and if these needs are met only to a limited degree by development of atomic power. Estimates by the UN (1955) indicate that during the first decade of the 21st century fossil fuel combustion could produce an amount of carbon dioxide equal to 20% of that now in the atmosphere (Table 1).* This is probably two orders of magnitude greater than the usual rate of carbon dioxide production from volcanoes, which on the average must be equal to the rate at which silicates are weathered to carbonates. Thus human beings are now carrying out a large scale geophysical experiment of a kind that could not have happened in the past nor be reproduced in the future. Within a few centuries we are returning to the atmosphere and oceans the concentrated organic carbon stored in sedimentary rocks

---

* World Production of CO_2 from the use of limestone for cement, fluxing stone and in other ways was about 1% of the total from fossil fuel combustion in 1950.

Table 1. CO_2 added to atmosphere by consumption of fossil fuels

Decade	Average amount added per decade Measured		Cumulative total added			
	10^{18} gms	% Atm CO_2	10^{18} gms	% Atm CO_2	10^{18} gms	% Atm CO_2
1860—69	0.0054	0.23	0.0054	0.23		
1870—79	0.0085	0.36	0.0139	0.59		
1880—89	0.0128	0.54	0.0267	1.13		
1890—99	0.0185	0.79	0.0452	1.92		
1900—09	0.0299	1.27	0.0751	3.19		
1910—19	0.0405	1.72	0.1156	4.91		
1920—29	0.0470	2.00	0.1626	6.91		
1930—39	0.0497	2.11	0.2123	9.02		
1940—49	0.0636	2.71	0.2759	11.73		
	Forecast					
	Assuming fossil fuels are used to meet future requirements of fuel and power as estimated by UN (1955)				Assuming fossil fuel consumption remains constant at estimated 1955 rate	
1950—59	0.091	3.9	0.367	15.6	0.367	15.6
1960—69	0.128	5.4	0.495	21.0	0.458	19.5
1970—79	0.176	7.5	0.671	28.5	0.549	23.4
1980—89	0.247	10.5	0.918	39.0	0.640	27.2
1990—99	0.340	14.5	1.258	53.5	0.731	31.1
2000—09	0.472	20.0	1.730	73.5	0.822	35.0

over hundreds of millions of years. This experiment, if adequately documented, may yield a far-reaching insight into the processes determining weather and climate. It therefore becomes of prime importance to attempt to determine the way in which carbon dioxide is partitioned between the atmosphere, the oceans, the biosphere and the lithosphere.

The carbon dioxide content of the atmosphere and ocean is presumably regulated over geologic times by the tendency toward thermodynamic equilibrium between silicates and carbonates and their respective free acids, silica and carbon dioxide (Urey, 1952). The atmosphere contains and probably has contained during geologic times considerably more CO_2 than the equilibrium concentration (Hutchinson, 1954), although uncertainties in the thermodynamic data are too great for an accurate quantitative comparison. Equilibrium is approached through rock weathering and marine sedimentation. Estimated rates of these processes give a very long time constant of the order of magnitude of 100,000 years. Rapid changes in the amount of carbon dioxide produced by volcanoes, in the state of the biosphere, or as in our case, in the rate of combustion of fossil fuels, may therefore cause considerable departures from average conditions.

The answer to the question whether or not the combustion of coal, petroleum and natural gas has increased the carbon dioxide concentration in the atmosphere depends in part upon the rate at which an excess amount of CO, in the atmosphere is absorbed by the oceans. The exchange rate of isotopically labeled CO_2 between atmosphere and ocean, which, in principle, could be deduced from C^{14} measurements, is not identical with the rate of absorption, but is related to it. . . .

SECULAR VARIATION OF CO₂ IN THE ATMOSPHERE

In the preceding section of this paper, two simplifying assumptions were made when estimating the exchange rate of CO_2 between the atmosphere and the oceans: (1) that the rate constants k_1 and k_2 were not affected by a small increase of the exchangeable carbon reservoir such as that from industrial fuel combustion, and (2) that except for that size increase, no other changes in the sizes of the oceanic and atmospheric carbon reservoirs have taken place. . . .

Because of the peculiar buffer mechanism of sea water, however, the increase in the partial CO_2 pressure is about 10 times higher than the increase in the total CO_2 concentration of sea water when CO_2 is added and the alkalinity remains constant. . . .

In contemplating the probably large increase in CO_2 production by fossil fuel combustion in coming decades we conclude that a total increase of 20 to 40% in atmospheric CO_2 can be anticipated. This should certainly be adequate to allow a determination of the effects, if any, of changes in atmospheric carbon dioxide on weather and climate throughout the earth.

ROGER REVELLE

TESTIMONY BEFORE THE HOUSE COMMITTEE ON APPROPRIATIONS, FEBRUARY 8, 1956

Historians focused on the history of climate science often point to the Revelle and Suess paper in *Tellus* as a preliminary warning on anthropogenic global warming, but Revelle did not confine his discussions of the great geophysical experiment to scientific papers. Here, in two hearings from 1956 and 1957 on the International Geophysical year—an eighteen-month program of geophysical research designed to promote international scientific cooperation and "friendly" Cold War scientific competition—Revelle uses the CO_2 issue as a way to highlight the need for more basic research into the oceans and the atmosphere. Revelle's congressional hosts—Sidney Yates and Albert Thomas, both Democrats—would have welcomed an opportunity to join their party colleagues in questioning President Eisenhower's commitment to science and technology in the face of the Soviet threat. How are Revelle's comments here similar to or different from his statements in the *Tellus* paper with Suess? What is at stake in this testimony?

• •

OCEANS AND SEAS

Dr. REVELLE. Mr. Chairman, as I sat listening to this, I have been thinking that this scene is not very much different than the scene

---

National Science Foundation: International Geophysical Year, Hearings before the House Committee on Appropriations, 84th Cong. (1956). Excerpt: pp. 465–66, 472–73.

when Columbus asked the sovereigns of Spain for money to explore the new world. There are Columbuses all over the world doing that.

Mr. THOMAS. There is nobody sovereign on this side.

Mr. YATES. I must say, I was shocked by Dr. Porter, though, when I was thinking of that very example. Columbus pointed out very dramatically that the earth was round. Dr. Porter denies it.

Dr. REVELLE. There are Columbuses all over the world asking their sovereigns for support for this program of exploration, really a new age of exploration which is just beginning. At least in oceanography, the Russian Columbuses have been doing a lot better than we have.

Mr. THOMAS. Tell us a little something about yourself here.

SCRIPPS INSTITUTION OF OCEANOGRAPHY

Dr. REVELLE. I am director of the Scripps Institution of Oceanography of the University of California, which I think Mr. Phillips will agree is one of the most important centers of oceanographic research in California.

Mr. PHILLIPS. I am also a biased witness.

Mr. THOMAS. How long have you been there, Doctor?

Dr. REVELLE. I have been there since 1931, off and on. I spent 7 years in the Navy in the middle thirties, but before and after the war I was at the Scripps Institution.

Mr. THOMAS. What did you do in the Navy?

Dr. REVELLE. Office of Naval Research, and with Admiral Blandy's task force at Bikini.

Mr. THOMAS. What is the nature of your work with the task force? You are awfully modest. We have a great American here and we are trying to get a little on the record here, Doctor.

RADIOACTIVITY MEASUREMENTS

Dr. REVELLE. What I actually did for the task force out at Bikini was to try to think of ways in which we could measure and observe and record the effects of the atomic explosion on the waters and on the fish and all the other organisms that live in that area and on the atolls, themselves.

We have quite a large group to do this and we have been doing it off and on ever since. This spring, for example, the Scripps Institution will have one of its big ships and about 75 of its people out at Einewetok in various aspects of this same program, particularly studying the effect of radioactivity.

Mr. THOMAS. Now, get back to your subject. Thank you, Doctor.

FUELS AND CARBON DIOXIDE

Dr. REVELLE.... There is still one more aspect of the oceanographic program which I thought you gentlemen would be interested in. This is a combination of meteorology and oceanography. Right now and during the past 50 years, we are burning, as you know, quite a bit of coal and oil and natural gas.

The rate at which we are burning this is increasing very rapidly. This burning of these fuels which were accumulated in the earth over hundreds of millions of years, and which we are burning up in a few generations, is producing tremendous quantities of carbon dioxide in the air. Based on figures given out by the United Nations, I would estimate that by the year 2010, we will have added something like 70 percent of the present atmospheric carbon dioxide to the atmosphere. This is an enormous quantity. It is like 1,700 billion tons. Now, nobody knows what this will do. Lots of people have supposed that it might actually cause a warming up of the atmospheric temperature and it may, in fact, cause a remarkable change in climate....

WARMING OF THE EARTH

We may actually, for example, find that the Arctic Ocean will become navigable and the coasts become a place where people can live, then the Russian Arctic coastline will be really quite free for shipping, as will our Alaskan coastline, if this possible increase in temperature really happens.

This would have the effect, if it does happen, of changing the character of the Russians as opposed to ourselves. We are now the greatest maritime nation on the earth. We are essentially living on an island surrounded by a world-circling ocean.

If the Russian coastline increases by something like 2,000 miles or so, the Russians will become a great maritime nation.

Mr. THOMAS. Didn't I read, from what Dr. Gould says, we have been warming up for the last 50 years?

Dr. REVELLE. The reason may be because of the carbon dioxide. It may be because we have been adding carbon dioxide to the atmosphere. There are two questions here that we have to answer.

One question is, How much of the carbon dioxide that we are producing goes into the ocean, and one of the aspects of this IGY oceanographic program is to try to find out what proportion of the total carbon dioxide produced by the burning of fossil fuels goes into the ocean and how much stays in the atmosphere.

We think that something like 30 percent of it stays in the atmosphere. This would be, in other words, by the year 2000 an increase of about 25 percent of the amount now in the atmosphere.

This may actually have the effect that I spoke of. Nobody really knows, because we don't know what the effect of the CO_2 in the atmosphere is. We can't do it by figuring it on paper. It has to be done by experiment.

Here we are making perhaps the greatest geophysical experiment in history, an experiment which could not be made in the past because we didn't have an industrial civilization and which will be impossible to make in the future because all the fossil fuels will be gone. All the coal and gas and oil will be used up.

In this 100-year period, we are conducting, in effect, this vast experiment, and we ought to adequately document it. One of the main parts of the oceanographic program is to try to do just that.

ROGER REVELLE

TESTIMONY BEFORE THE HOUSE COMMITTEE ON APPROPRIATIONS, MAY 1, 1957

••

STUDIES OF THE EARTH'S HEAT BALANCE

Dr. REVELLE. I think the best way to introduce this subject is to point out to you gentlemen something which is not often thought of, and that is that the earth itself is a space ship. We were talking this morning about the beginning of space travel, about the beginning of man's leaving the earth and going to the stars, but, in fact, we have lived here on this space ship of our earth for a good many hundred thousand years, and we human beings are specifically adapted to it. We are built to take advantage of the earth. Our whole physiology and psychology really depend upon the characteristics of the earth. This is something, of course, that we do not know about, or did not know about when we human beings were developing. Until recently people did not even know the earth was round.

Mr. THOMAS. You are talking like an environmentalist. I thought that you believed in heredity.

Dr. REVELLE. We are certainly shaped by the earth on which we live. A simple example is the fact that we breathe oxygen. This planet of ours is the only planet in the solar system that we know of that has free oxygen on it.

National Science Foundation: Report on International Geophysical Year: Hearings before the House Committee on Appropriations, 85th Cong. (1957). Excerpt: pp. 104–8.

Mr. THOMAS. That will not be free long. The Federal Government will tax it.

Dr. REVELLE. That has very definitely determined our physiology, the fact that there is oxygen and the fact that there is a great deal of water on the earth. Again, this is the only planet that has large bodies of liquid water on it. The fact that water has such a great capacity for storing heat means that it can absorb a great deal of radiation and not change its temperature very much.

Mr. THOMAS. How do we know that these other planets do not have any nice warm springs and cold springs and a few little creeks?

Dr. REVELLE. At least the astronomers have never been able to find the evidence. I personally would like to go to Mars and find out for myself. I think the evidence is fairly good that there is a very small amount of water there.

Mr. THOMAS. Has it been proven there is not?

Dr. REVELLE. One simple piece of evidence is this: On Mars there is a kind of frost cover around the North Pole, but the atmosphere on Mars is so dry that this ice cap, if you will, actually disappears every summer and moves through this very dry air and settles again on the South Pole of Mars. So the ice cap on Mars is quite different from ours. It actually migrates from pole to pole during the Martian year, and this shows that the atmosphere at Mars is dry as dust, drier than it has recently been in Texas.

Mr. THOMAS. You have announced a very fundamental principle of weather when you said that these vast bodies of water are a reservoir for tremendous heat loads.

Dr. REVELLE. That is correct, sir, and you are obviously going to push me as to what we are going to study. I will come to that right now.

What we are concerned about in the meteorological, the oceanographic, and the glaciological programs is what happens on the earth because of the radiation coming from the sun.

In other parts of the IGY program the nature of this radiation, what actually comes into the earth, will be studied. The next question is, What will the earth do with it? It makes climate, the ocean currents and the ocean circulation, which provide a fertile area for the fisheries.

HISTORY OF CLIMATE CHANGES

Now, the interesting and curious question is this: The amount of energy coming in from the sun each year can be fairly accurately calculated, but nobody knows how much goes back. In principle, if nothing was changing on the earth, just as much heat would go away from the earth as comes in. In fact, this is generally assumed to be the case. We know that in many times in the past it has not been the case. For example, we know that about 10,000 years ago the earth was a lot colder than it is now and that a lot of heat was locked up in the ice.

Quite abruptly, about 10,000 years ago, the earth apparently heated up quite a bit. A great portion of the icecaps melted, perhaps in a rather short space of time, and a few thousand years later the climate was actually considerably warmer than it is today. This happened several times in the relatively recent past from the standpoint of the geologists; that is, within the last few hundred thousand years.

There were at least four occasions when ice covered a large part of the Northern Hemisphere. There were great icecaps coming down well below Chicago and over most of northern Europe and over a good part of Asia.

This is perhaps the most dramatic example we have of the fact that the climate of the earth changes very markedly from time to time, and a long range question which is often talked about is this: Will the ice come down again? Will we enter into a new dark age of ice? I do not think this is a very practical question for people living in this generation, but the shorter time climatic changes, it seems to me, are of great importance.

Let me give you an example of my own experience. In California we are very much concerned about our water supply. We do not have enough water. Southern California is just barely livable at best. Our population is growing so fast that we are trying to get water from the northern part of the State and perhaps farther north, and of course from the Mountain States where the Colorado River originates.

Now, the people in the northern part of the State and those other States are very much concerned about this; they are afraid that perhaps at some time in the future they will not have enough water. They have too much now, but they may end up not having enough. This could very easily happen if we had a rather slight climatic change over the next 50 to 100 years.

FORECASTING CLIMATE

President Eisenhower recently said, and I quite agree, that water is perhaps becoming the most important resource of the United States. One of the fundamental parts of this water supply problem is the possibility of the water supply of the United States becoming slightly greater or slightly less as a result of climatic changes. The meteorologists have gradually developed during the past 50 to 100 years a fairly good ability to forecast weather. Dr. Wexler can tell you, if he looks at a map, about 80 percent of the time what the weather is going to be tomorrow—whether we are going to have a rainstorm or a thunderstorm, or, early in the season, a snowstorm. No meteorologist has the least idea how to forecast climate.

If you ask a meteorologist what the average rainfall or temperature will be 10 years from now, let alone a hundred years from now, he would not be able to tell you in the slightest degree. Yet the forecasting of climate is in fact, I think, a more important job—more important to the future of our own species than the forecasting of weather.

The problem is somehow to develop a means of forecasting.
Mr. THOMAS. You used the word "climate" there for the long range, and "weather" for the day-to-day basis?
Dr. REVELLE. That is right.

EFFECTS OF FOSSIL FUELS ON CLIMATE

The last time that I was here I talked about the responsibility of climatic changes due to the changing carbon dioxide content of the atmosphere and you will remember that I mentioned the fact that during the last 100 years there apparently has been a slight increase in the carbon dioxide because of the burning of coal and oil and natural gas.

If we look at the probable amounts of these substances that will be burned in the future, it is fairly easy to predict that the carbon dioxide content of the atmosphere could easily increase by about 20 percent. This might, in fact, make a considerable change in the climate. It would mean that the lines of equal temperature on the earth would move north and the lines of equal rainfall would move north and that southern California and a good part of Texas, instead of being just barely livable as they are now, would become real deserts.

Mr. THOMAS. Why now?

Dr. REVELLE. Because the precipitation would be considerably less than now.

The problem here is that the carbon dioxide seems to absorb—we think it does absorb—the so-called infrared radiation, the radiation going back into space, away from the earth. In order to get as much heat out as comes in, you have to increase the temperature of the earth by a relatively small amount, about 1½ degrees.

There is what the electronics people would call feedback. . . .

There are many theories. It is awfully hard to make a choice between them or to know how to test them. This carbon dioxide thing that I was talking about is in fact a way to test some of these theories.

Mr. THOMAS. You did not spell out very clearly your carbon dioxide theory. Spell it out again. Carbon dioxide absorbs the infrared rays that are bouncing back from the earth, and when they are absorbed, that absorbs the heat and therefore what?

Dr. REVELLE. It raises the temperature. It is like a greenhouse. When you go into a greenhouse the temperature is higher than it is outside. The reason is because the glass lets the light in, the visible light. It will not let the heat out, so the temperature in the greenhouse has to rise until you have a balance, so that the heat escaping is equal to the energy coming in in the form of light. This may happen in the atmosphere; in fact, we know it does with water vapor. We know the atmosphere around the earth makes the earth a gigantic greenhouse and this is why: instead of having about minus 100° on the cold side and plus 200° on the warm side, such as you will have on the satellite, we have a fairly equitable temperature between day and night. The earth's atmosphere acts like an insulating blanket. . . .

Only God knows whether what I am saying is true or not. What I am driving at is that this business of the carbon dioxide production is in fact a way of studying climatic changes.

HOWARD T. ORVILLE

THE IMPACT OF WEATHER CONTROL ON THE COLD WAR

By early 1958, well into the IGY and just months after the Soviets launched the satellite *Sputnik,* support for American science and technology had become a major national issue and a point of attack for ambitious congressional Democrats. Here, retired Navy captain Howard T. Orville enumerates the possibilities for weaponized weather and climate control in a Cold War context within a larger hearing on U.S. research into weather modification. Within both scientific and political circles, the idea was that basic research on climatic processes—including the impacts of CO_2, which Orville cites as unintentional climate modification—might lead to better weather and climate prediction; better weather and climate prediction might lead to effective and useful weather and climate control.*

. . . .

As much as we dislike it, today we are engaged in a cold war with a nation and its satellites whose leaders are stark realists and who will stop at nothing to achieve their objective—the absolute domination of communism throughout the world. President Eisenhower, in his recent

---

Howard T. Orville, "The Impact of Weather Control on the Cold War," in *Weather Modification Research: Hearings of the House Committee on Interstate and Foreign Commerce,* 85th Cong. (1958). Excerpt: pp. 51–53.

* For more on weaponizing the environment during the Cold War, see Jacob Darwin Hamblin, *Arming Mother Nature: The Birth of Catastrophic Environmentalism* (Oxford: Oxford University Press, 2013).

state of the Union message, stated, "The threat to our safety and to the hope of a peaceful world is . . . Communist imperialism."

One way of achieving this goal is by means of a worldwide propaganda campaign to lower the stature of the United States in the eyes of the Western Powers. It may well be assumed that they are constantly searching for new sources of material in waging this campaign.

Only last week when the Army successfully fired the Explorer were we able to partially recover some prestige from the terrific loss sustained when the Russians fired two satellites into space. One is still orbiting around the earth. This experience should have taught us that we must be constantly on the alert for seeking the propaganda advantage over the Soviets.

If you will think back to September 1957 or earlier, you will recall that perhaps not one person in a million would have predicted the successful launching of a satellite that would orbit the globe for months. Fantastic! Unheard of! Impossible!! would have been some of the remarks received.

This, then, is an experience very fresh in our minds and should make us aware of the fact that the impossible may become a reality long before the best scientific or military brains can even envisage such an eventuality. In the past we have been slow in assessing the propaganda value of scientific achievements. Let's hope that the future will see us more clairvoyant.

This brings me to the subject of the discussion today—the impact of weather control on the cold war. Few areas of science have implications so profound to all mankind as the study of the atmosphere and the phenomena which occur in it.

Perhaps it would be appropriate to define the term "weather control." Weather control means that our knowledge of atmospheric processes has reached the level where we are able to apply manmade techniques to large scale weather patterns to start a chain reaction that will produce known results over a specific portion of the globe for a known period of time. For example, if an unfriendly nation were able to bring about the recurring destructive cold spells that have plagued Florida for the past 2 months, this would be absolute weather control.

Prior to the sputnik era the best scientific opinion was that the day of weather control was perhaps 30 to 60 years away, maybe even longer.

When, in 1954, I suggested that with the expenditure of $11 billion over a period of 30 to 40 years we might achieve complete weather control, I was severely criticized by my contemporaries.

Since October 1957 there have been many statements (some far too optimistic) by scientists of world renown. Probably one of the most significant statements was made by Dr. Edward Teller before the Senate Military Preparedness Committee. He stated that he was "more confident of getting to the moon than changing the weather, but the latter is a possibility. I would not be surprised if (the Russians) accomplished it in 5 years or failed to do it in the next 50."

What are some of the methods of achieving weather control? A number of methods have been suggested for bringing about major influences on weather or climate. To date none of them can be called effective methods of weather control because, if applied by any nation, no one knows what would be the resulting action of such drastic intervention on our own country or that of other nations. Until we are able to predict with certainty what will happen when manmade methods are applied, weather control should not be attempted.

USE OF COLORED PIGMENTS (LAMP BLACK)

One method of controlling the weather that has often been mentioned is the large scale use of colored pigments over the polar ice surfaces. It is well known that the persistence of large ice fields is due in part to the fact that ice both reflects sunlight energy and radiates away terrestrial energy at even higher rates than land surfaces. If microscopic layers of colored matter were spread on the ice or in the air above it, it would alter the reflection-radiation balance, melt the ice, and change the local climate. Measures that would bring about such a change are technically and economically feasible.

WARMING CAUSED BY INCREASED CARBON DIOXIDE CONTENT

The carbon dioxide that is released to the atmosphere by industry in the burning of coal and oil and their derivatives (most of it during the last generation) may have changed the composition of the atmosphere sufficiently to bring about a general warming of the world's temperature by about 1° Fahrenheit. This warming represents a 2 percent increase in the carbon dioxide content of the air. Studies show that when the amount has increased to 10 percent the icecaps will begin to melt with the resulting rise in the sea level. Coastal cities and areas such as New York and Holland would be inundated.

COOLING CAUSED BY RADIOACTIVE "DUST"

When the volcano Krakatoa erupted in 1883 it released an amount of energy by no means exorbitant. If the dust of the eruption had continued to circle the globe at high altitudes for 15 years instead of only 3 years. it would have lowered the world's temperature by 6°. Five such eruptions would have brought another Ice Age. This temperature is only about 15° below the present world's temperature.

Dr. Teller has testified in congressional hearings that if 2,000 H-bombs were to be exploded over a period of 20 years, the "dust" floating around in the stratosphere would be sufficient to cause a cooling to bring on a new Ice Age. This, of course, discounts the possibility that the "dust" might serve as freezing nuclei.

ARTIFICIAL SATELLITES AS PLATFORMS FOR WEATHER CONTROL

Weather satellites equipped with powerful telescopes and television attachments could chart cloud movements, detect the birth of hurricanes or other severe storms, and vastly improve our surveillance of weather patterns.

Dr. Hermann Oberth, of Germany, foresees a gigantic mirror "hung" in space. It would focus the sun's rays as [a] giant magnifying glass at any desired intensity and beam. The sun's rays could light entire cities or other areas safely at night. The heat of the rays might be used to prevent killing frosts over orchards, or melt Atlantic icebergs, open frozen harbors, and probably bring about artificial control of the weather.

Senator Lyndon Johnson of Texas stated recently: "From space, the masters of infinity could have the power to control the earth's weather, to cause drought and flood, to change the tides and raise the level of the sea, to divert the Gulf Stream, and change temperate climates to frigid."

I have read that at least twice in the past 12 months Moscow has boasted of large public works projects that would upset the entire wind circulation pattern of the Northern Hemisphere; conducted numerous nonpublicized but still detectable experiments apparently aimed at finding ways to speed the melting of polar icecaps; and has even offered to join the United States in a project to turn the Arctic Ocean into a sort of warm water lake by melting the polar icecap.

Russia's apparent preoccupation with plans to change the climate of the Northern Hemisphere is easy to understand. The Soviets and their satellites stand to profit more, from an economic standpoint, than any other combination of nations.

The proposal to turn the now frozen Arctic Ocean into a warm water lake came over Moscow radio just 3 days after last Christmas.

A Russian engineer, Arkady Borisovich Markin, proposed that a team of international scientists cooperate in designing a dam to redirect the waters of the Pacific Ocean to relieve the severe cold of the Northern Hemisphere. Such a dam built across the Bering Straits would be fitted with thousands of nuclear energy powered pumps that would pump the warmer waters from the Pacific into the colder Arctic Ocean. At other times the pumps on the Bering Straits would pump the water from the Arctic Ocean to the Pacific, canceling out the Greenland, Labrador, and other cold oceanic streams.

Just how such a plan would work is not clear to me; but if, as Markin states, the dam would raise the temperatures of such cities of New York, London, Berlin, Stockholm, and Vladivostok 11° to 14½°, then the melting of the icecaps and the release of landlocked waters would cause seacoast towns and cities to be flooded by the expected rise in the sea levels. This means that Russia might then enjoy warm water ports and mild temperatures similar to our South Atlantic States, but at the expense of great flooding of the Western nations' coastal areas.

Other great schemes for Russian public works programs to divert the course of several of its rivers from the Arctic Ocean to the Caspian Sea are less dramatic, but might eventually upset the present weather patterns over Siberia, which in time would affect the weather around the Northern Hemisphere.

There are many other methods by which effective weather control may be practiced by an unfriendly nation, but they are too numerous to even mention by name.

A moment's consideration of any of the methods mentioned here today should serve to convince any skeptic that weather control can have frightening and disastrous consequences if any unfriendly nation succeeds in gaining a break-through before we do.

This is, I think, what Dr. Teller meant when he recently stated:

"The Russians can conquer us without fighting through a growing scientific and technological preponderance. The Russians may advance so fast in science and leave us so far behind that their way of doing things will be "the way," and there will be nothing we can do about it. Imagine, for instance, a world in which the Russians can control weather on a big scale, where they can change the rainfall over Russia. This might very well influence the rainfall in our country in an adverse manner. What kind of world will it be where they have this new kind of control and we do not?"

Now, what courses of action can we take to get out of our present extremely vulnerable position for future Soviet propaganda attacks in the field of weather control? There are two important steps that must be taken.

First, see that Senate bill S.86 is passed. Fortunately, through the vision and courage of several Members of Congress, notably, Senator Francis Case, of South Dakota; Senator Clinton Anderson, of New Mexico; Senator Arthur Watkins, of Utah; and Senator Warren Magnuson, of Washington, there is now pending in Congress a bill, S.86, to set up a permanent research program under the National Science Foundation for an accelerated study in all aspects of atmospheric physics. When this bill, which was originated by Senator Francis Case, is passed and is aggressively implemented it will provide for the urgently needed program in basic and applied research.

Second, there must be an awareness and an appreciation of the valuable service that the United States Weather Bureau is performing today under most stringent circumstances. It is understaffed; much of its equipment is obsolete; research facilities are practically nonexistent; its research program is far below that which it should be during these critical times; and the prospects for getting the funds needed to correct these deficiencies are very poor. Other weather services of the Defense Department dealing primarily with military implications of weather control should be given [the] strongest possible support.

Since weather control is an ideal tool for waging the relentless cold war against the Western Powers, we must not become complacent and we must not be caught as we were when Sputnik I was launched last October.

We must take seriously the truth of Vice President Nixon's statement, "the Kremlin has reaffirmed its goal of world domination by nonmilitary means if possible, and by war if necessary."

Weather control has many important nonmilitary applications and just as many military uses.

NATIONAL SCIENCE FOUNDATION

PRELIMINARY PLANS FOR A NATIONAL CENTER FOR ATMOSPHERIC RESEARCH: PROGRESS REPORT OF THE UNIVERSITY COMMITTEE ON ATMOSPHERIC RESEARCH

As early as 1956—amid planning for the IGY—a number of prominent atmospheric scientists began to call for the consolidation of research on the earth's atmosphere then being conducted across a wide array of universities and government institutions. Championed in part by National Science Foundation president Detlev Bronk, the result was a National Institute for Atmospheric Research (NIAR, later renamed the National Center for Atmospheric Research, NCAR) in Boulder, Colorado. The institute would be administered by a consortium of universities through the University Corporation for Atmospheric Research (UCAR), insulating it from direct government or military oversight in ways that agencies like NASA were not. The founding document, commonly referred to as the "Blue Book," outlined the scientific objectives of the institution—along with its often not subtle Cold War influences. Where do the concerns that Revelle, Orville, and their congressional hosts articulate appear in the Blue Book, and how does the form and purpose of the Blue Book constrain those concerns?*

National Science Foundation, *Preliminary Plans for a National Center for Atmospheric Research: Second Progress Report of the University Committee on Atmospheric Research* (Washington, DC: National Science Foundation, 1959). Excerpt: Cover letter, pp. iii–iv, vii–ix, 26–28, 44–46.

* For more on the founding of NCAR, see Joshua P. Howe, *Behind the Curve: Science and the Politics of Global Warming* (Seattle: University of Washington Press, 2014), 27–32.

Dr. Alan T. Waterman, Director
National Science Foundation
Washington 25, D. C.

Dear Dr. Waterman:

I have the honor to transmit herewith, on behalf of the University Committee on Atmospheric Research, our Second Progress Report which has been prepared with the support of grant G5807 from the National Science Foundation.

This report is based on an analysis of the scientific problems posed by the atmosphere and of the facilities needed to mount an effective attack on them. It represents a distillation of the opinions of some one hundred and fifty research scientists, perhaps half of all those engaged in basic atmospheric research in this country. It is striking that, almost without exception, the working scientists agreed that an effective attack on the major problems of the atmosphere requires a National Institute for Atmospheric Research of the sort described in the report.

Although the report is concerned primarily with the scientific concept of a National Institute it seemed necessary to translate this into a physical plan and from this to make reasonable estimates of costs. This has been done with as much care and accuracy as possible. On the other hand we consider it essential that the actual research program, facilities and buildings be planned by the scientists who will staff the Institute. Such actual plans will differ in detail from those presented in the report but we are confident that the broad concept will remain valid.

We are convinced of the scientific need for a National Institute and firmly believe that the benefits that will eventually arise from a more adequate basic understanding of our atmospheric environment more than justify the expenditure of Federal funds for this purpose. We will await with great interest your reactions to this report.

Sincerely Yours,

Henry G. Houghton
Chairman

PREFACE

This report summarizes the activity and the studies that have taken place over the last year and a half relative to the establishment of a National Institute for Atmospheric Research to be operated by a group of universities and supported by the National Science Foundation. It was first proposed by the Committee on Meteorology of the National Academy of Sciences. Subsequent studies by the University Committee on Atmospheric Research (UCAR), representing fourteen academic institutions with interest in the atmospheric sciences, confirmed the need for the Institute. A series of seventeen research planning conferences was held under the auspices of UCAR with support provided by the National Science Foundation. These conferences covered nearly all aspects of atmospheric research and were attended by some one hundred and fifty scientists familiar with one or more of these aspects.

From the meetings of UCAR and the research planning conferences there has emerged a consensus of the nature of the Institute, the broad fundamental scientific problems of the atmosphere to which it should address itself, and the nature and kind of research facilities which it should provide. From these ideas, tentative plans for the orderly establishment of the Institute over a six-year period have been formulated and estimates made of the budgetary support that will be required. . . .

SUMMARY

A National Institute for Atmospheric Research has been recommended and supported by several representative groups in the scientific community as an essential part of an over-all expansion in atmospheric research in this country. This Institute will be devoted exclusively to basic research and will be operated by a group of universities with primary support from the National Science Foundation. Its principal function will be to enhance the effectiveness of atmospheric research scientists and to bring to bear on the problems of the atmosphere the full competence of the scientific community. It will do this by fostering the vigorous interdisciplinary effort and providing the complex of large facilities and technological support that are required to make significant advances in describing the salient features of the atmosphere and in understanding the important physical processes which determine its behavior.

There are four compelling reasons for establishing a National Institute for Atmospheric Research:

1. The need to mount an attack on the fundamental atmospheric problems on a scale commensurate with their global nature and importance.
2. The fact that the extent of such an attack requires facilities and technological assistance beyond those that can properly be made available at individual universities.
3. The fact that the difficulties of the problems are such that they require the best talents from various disciplines to be applied to them in a coordinated fashion, on a scale not feasible in a university department.
4. The fact that such an Institute offers the possibility of preserving the natural alliance of research and education without unbalancing the university programs.

The scientific program of the Institute will be focused on the fundamental problems in four principal areas of research:

1. Atmospheric motion.
2. Energy exchange processes in the atmosphere.
3. Water substance in the atmosphere,
4. Physical phenomena in the atmosphere.

Research activity at the Institute will be grouped around three major divisions: Physical Research (including instrumentation), Mathematical Research, and Chemical Research. Scientists from each of these divisions will be concerned with more than one of the principal areas of research listed above. Major facilities at the Institute will include (1) a group of instrumented airplanes consisting of two light twin-engine planes, two medium twin-engine planes, three DC-6's and two B-57's, (2) a scientific library, (3) a large-scale transistorized electronic computer, (4) a spectroscopic laboratory, (5) microwave and optical radars, together with sferics equipment and infrared equipment for probing the atmosphere with electromagnetic radiation and (6) well equipped electronics and machine shops. Housing for temporary staff is required.

The estimated personnel requirements for the research program contemplated at the Institute include 108 scientists, 206 technical and professional people and 236 support people. It is further estimated that six years will be required to obtain a full staff of the competence that is desired. The scientific manpower requirements do not present insurmountable problems. About one-half of the scientists will be individuals with training in the atmospheric sciences and about one-half will be drawn from the parent disciplines of physics, mathematics and chemistry, and from engineering. Approximately one half of the staff will have permanent status and the balance will be on a visiting status. The facilities of the Institute will be open to competent scientists from all disciplines without regard for institutional or organizational affiliation. Budgetary support of approximately $1,141,000 is estimated as the requirement for Fiscal Year 1960. The yearly budget will gradually increase toward an annual figure between $14,000,000 and $15,000,000 by Fiscal Year 1965. On the basis of preliminary plans for the six-year period from Fiscal Year 1960 to Fiscal Year 1965, inclusive, the total capital costs are estimated to be $33,253,000. On the same basis the total operating expenses are estimated to be $38,042,000. . . .

Many of the fundamental problems of the atmosphere are not likely to be solved without sustained effort. Substantial progress toward a satisfactory understanding of the atmosphere will probably have to be measured in decades rather than in years. It is essential that support for the activities of the Institute be of a stable and longterm nature even though some problems are susceptible of immediate solution given the proper scale of effort. Because the research emphasis will be on basic problems and particularly those beyond the capacities of individual university departments it seems necessary that the principal source of support for the Institute must come from the Federal Government, although support from private and industrial sources will be encouraged. Since the research to be undertaken is basic in nature, it is desirable that federal support come through the National Science Foundation rather than through operational agencies of the government. It is strongly urged that the Institute not serve as a channel for the disposition of federal funds toward the research efforts of the universities.

The central effort of the Institute must concern the advancement of knowledge of the whole atmosphere. This requires the attention of men of the highest competence who are trained in a wide variety of disciplines. The full breadth of such an enterprise can only be partially stated at the present time, but it is the point from which the Institute program must make its departure....

It remains to be established that proper justification exists for the Institute as it is now conceived. The consensus of each of the groups which has studied the problem—the Academy Committee, the University Committee on Atmospheric Research, the university administrations, the community of atmospheric scientists—is that the need for the Institute is urgent. The reasons were succinctly stated in the First Progress Report of UCAR (Appendix E). The national interest is clearly involved on three counts: (1) it is almost a national scandal that weather continues to take its yearly toll of life, wealth, and property without a greater effort to understand the physical processes that govern the atmosphere, (2) the possibility, slight though it may be, that one day a better understanding of the atmosphere might lead to effective large-scale weather control, and (3) the broad implications of the friendly yet thoughtful warning by Dr. Charles H. Malik, President, United Nations General Assembly, in an address at Harvard in September, 1958,

> ... "in the field of scientific and technological research there is a most serious competition with the West. The challenge here appears to be that unless the effort at the formation of scientists and technicians and at the promotion of pure and applied science is considerably intensified, the West is heading toward an inferior position in the foreseeable future"...

Funds for a research enterprise, as large as the Institute should be to achieve critical size, are highly unlikely from private sources....

Over and above their fundamental scientific importance, there are three reasons why provision should be made for inclusion of problems in this general area in the scientific program of the Institute. The first is, as pointed out in the conference on the upper atmosphere, the worldwide activities of the International Geophysical Year have resulted in the amassing of an astounding volume of extremely valuable scientific data. The reduction, analysis and interpretation of these data are going to require the competence of a strong interdisciplinary group equipped with excellent facilities for large-scale data processing. While it would be highly undesirable to concentrate all of the effort in analyzing IGY data in one place (and would not be possible even if it were attempted), the Institute, as an international intellectual center with a marked interdisciplinary flavor and a first-rate computational laboratory, could serve as a focal point for bringing together and coordinating the several disciplines involved and make an important contribution to the effective utilization of these data—many of which are concerned with atmospheric phenomena which fall into the general category under discussion. Secondly, man's activities in consuming fossil fuels during the past hundred years, and in detonating nuclear weapons during the past decade have been on a scale sufficient to make it worthwhile to examine the effects these activities have had upon the atmosphere. Reference is made here to the still unsolved question of whether the carbon dioxide content of the atmosphere is increasing as a result of combustion processes and the even more elusive question as to possible changes in the earth's electrical field as a result of nuclear explosions. Thirdly, by nature most of these problems fall exactly into the realm of basic research and many of them require facilities on the scale envisioned for the Institute as well as an interdisciplinary approach.

PART 3

MAKING GLOBAL WARMING GREEN

Between Revelle and Suess's 1957 *Tellus* paper and the advent of the Intergovernmental Panel on Climate Change in the late 1980s, climate change moved from the fringes of Cold War research to the center of American and international environmental politics. That transition was episodic, contingent, and often highly problematic, however, and the challenges of framing CO_2-induced warming as an environmental issue—making global warming green—unfolded in a variety of changing historical, institutional, and political contexts.

From very early on in this story, scientists understood that increased atmospheric CO_2 presented a unique and uniquely challenging type of environmental threat, certainly different in scale, but also different in kind from mainstream issues like air and water pollution, toxic wastes, or land use change, which became the focus of a national environmental movement in the 1960s. Beginning with Charles David Keeling's 1963 report for the Conservation Foundation—in which he defined CO_2 as a form of pollution—the documents here highlight the challenges of making a highly technical, largely intangible, long-term global problem mean something to constituencies in environmental organizations, participants in geopolitical negotiations, and the larger American public. They also highlight the sometimes awkward overlaps between scientists and environmentalists on global environmental issues. In 1965, CO_2-induced warming landed on President Lyndon Johnson's desk in a report titled *Restoring the Quality of Our Environment*, but the report came from the President's Science Advisory Council, and most of the recommendations called for more research rather than the kinds of legislation and immediate change environmentalists sought on issues like pesticides and air pollution.[1] Where do we draw the line between science and environmental activism on CO_2 in these decades, and how are the structure and tone of the documents both similar and different? How did the convergence of these two overlapping communities—scientific and environmental—change the discourse about climate change between the early 1960s and the early 1980s?

In the 1970s and 1980s, two national organizations—the American Association for the Advancement of Science (AAAS), the nation's largest scientific association; and the Sierra Club, the nation's largest environmental advocacy group—stood at the forefront of the process of making global warming a meaningful environmental issue. The AAAS, steeped in a tradition of science advocacy focused on increasing and maintaining funding for and public interest in scientific pursuits, sought to promote more and better science on climate change, relying on the "forcing function of knowledge" to drive good climate change policy in government agencies. In doing so, the organization gained remarkable access to government officials, especially under the Carter administration. Based on a joint meeting between the Department of Energy and the AAAS, David Slade of the Department of Energy actually constructed a proposal for incorporating climate change into DoE energy planning during Carter's theoretical second term.

Environmental organizations like the Sierra Club, meanwhile, came to the issue of climate change much more slowly and cautiously. Here again, as in the early history of climate science presented in the first chapter, the lens of the present can distort our understanding of the past. From the perspective of the twenty-first-century present, in which American environmental organizations have made climate change their central and defining challenge, it is easy to forget that the issue presented environmentalists with an unwieldy and largely unwelcome set of problems in the 1970s and 1980s. Few documents highlight these difficulties as clearly as Sierra Club director Michael McCloskey's memo outlining the types of issues the Sierra Club should take on internationally in the 1980s. How well does climate change fit with McCloskey's 1982 rubric for successful environmental campaigns? Why would the Sierra Club focus on "backyard" issues in the early 1980s rather than the prevalent global issues we worry about today? By the middle of the 1980s, the Sierra Club and other environmental organizations like Friends of the Earth had come to recognize global warming as an important environmental issue—and, in fact, Sierra Club international program director Patricia Scharlin's Venn diagram from the club's International Program Archives in 1977 reveals that some in the club had already begun to perceive climate and climatic change as central to the larger international picture of global environmental problems. As a whole, however, throughout the 1980s, environmental organizations still struggled to frame the issue as a pressing problem for constituencies interested in local, tangible environmental problems.

The differences between scientists' and environmentalists' responses to climate change between 1960 and 1985 pose something of a puzzle. Why would the same highly technical, deeply scientific characteristics that made global warming an ideal issue for AAAS involvement make it difficult for the Sierra Club to incorporate climate change into its larger program of environmental advocacy?

One answer lies in the institutions themselves. Like most of the archival work that makes up the heart of historical scholarship, finding, reading, and interpreting documents in institutional archives like those of the AAAS and Sierra Club requires creativity and diligence. But institutional records in particular also require a certain attention to organizational structure. Often—as is the case with the Sierra Club, where separate file collections exist for executive documents, program files, and membership documents—the structure of an archive directly reflects the structure of the organization. Sometimes, as in an organization like AAAS that relies heavily on the work of nonstaff committees, these structures are more fluid and opaque. In both cases, however, thinking about the hierarchies, divisions of labor, and forms of accountability in an organization can help make sense of the documents it produces. Who participates in setting organizational priorities? Who are the organization's constituents? And how do these things impact their approach to a problem like climate change?

The flowchart from David Slade's proposal for incorporating climate change into DoE energy planning in the 1980s further underscores the point, this time highlighting institutional differences not between two advocacy groups, but between the AAAS and a government agency. As an AAAS document, the flowchart represented a great success for one of the organization's programmatic foci in the late 1970s. In the larger context of the DoE's mission to oversee everything from the development of synthetic fuels to nuclear weapons intelligence and counterintelligence, however, Slade's flowchart was barely a blip on the radar. How do differences in these two institutions yield such differing interpretations of the same document?

Let's begin with the AAAS. If part of the AAAS's mandate was (and still is) to facilitate the use of good scientific information in making policy, why might the AAAS have seen in this flowchart a great success for its efforts to promote climate science in the late 1970s? What elements of the image best fit the AAAS's institutional objectives?

Next, think about the structure of the Department of Energy. The DoE—a new cabinet level agency created in 1977—oversaw the nation's energy resources at a large scale, including everything from hydroelectric generation to fuel security to nuclear power to waste disposal and energy conservation to nuclear weapons development. Slade was director of the DoE's relatively low-level Office of Health and Environmental Research in a period marked by a foreign-policy-related energy crisis involving Iran and a domestic shift toward coal-based synthetic fuels, all of which landed on the DoE's agenda. How does the DoE's mandate help put Slade and his chart in context? In terms of policy, what does Slade's framework for CO_2 actually call for? What kind of DoE institutional constraints has he embedded in the image, and why?

We can never know whether Carter would have incorporated CO_2 into his second-term energy agenda, but a document that looks quite hopeful in the context of the AAAS looks perhaps less convincing in the structural framework of the Department of Energy. In the end, Carter's second term was not to be; the Reagan administration eventually replaced Slade with a political acolyte with only limited interest in climate change, and the AAAS lost its pipeline into government. (Slade's wry note to AAAS's David Burns on the transition humorously underscores the way a change in administration changed the institutional priorities of the DoE, reinforcing the importance of institutions in thinking historically about climate change.) In the 1980s, scientists began to turn toward environmental organizations like McCloskey's Sierra Club to help reintroduce CO_2 to the public in yet a new historical context.[2]

NOTES

1. Four years later, Richard Nixon aide Daniel Patrick Moynihan brought the issue of climate change before another administration in a letter to John D. Erhlichman dated September 17, 1969. Nixon Presidential Library, www.nixonlibrary.gov/virtuallibrary/documents/jul10/56.pdf.

2. For more on the AAAS-DoE program and the Sierra Club's changing priorities, see Joshua P. Howe, "Climate, the Environment, and Scientific Activism," in *Behind the Curve: Science and the Politics of Global Warming* (Seattle: University of Washington Press, 2014); for more on the transition from Carter to Reagan, see Samuel P. Hays, "The Anti-Environmental Revolution," in *Beauty, Health, and Permanence: Environmental Politics in the United States, 1955–1985* (New York: Cambridge University Press, 1987), 491–526.

THE CONSERVATION FOUNDATION

IMPLICATIONS OF THE RISING CARBON DIOXIDE CONTENT OF THE ATMOSPHERE

A Statement of Trends and Implications of Carbon Dioxide Research Reviewed at a Conference of Scientists

One of Roger Revelle's goals in lobbying Congress for funds to monitor atmospheric CO_2 in the late 1950s was to create the infrastructure to document the "large-scale geophysical experiment" as it unfolded over time. To that end, he hired a young geochemist named Charles David Keeling to set up CO_2 monitoring stations at Mauna Loa, Hawaii, and elsewhere to create a set of global baseline measurements of carbon dioxide. For Keeling, however, the "Keeling Curve"—the upward-sloping, saw-toothed graph of atmospheric CO_2—had obvious environmental implications, and by the early 1960s he had begun to frame CO_2 in terms of the buzzword of early 1960s environmentalism: pollution. In 1963, working with a nonprofit group called the Conservation Foundation, Keeling made the case for thinking about CO_2 as a dangerous pollutant—one of the first articulations of the problem of understanding atmospheric CO_2 as an environmental issue. For Keeling and his Conservation Foundation coauthors, what problems did this early application of the pollution framework pose for talking about climate change in the 1960s?

· ·

Conservation Foundation, *Implications of Rising Carbon Dioxide Content of the Atmosphere: A Statement of Trends and Implications of Carbon Dioxide Research Reviewed at a Conference of Scientists* (New York: Conservation Foundation, 1963).

FOREWORD

On March 12, 1963, The Conservation Foundation assembled a conference of scholars to discuss the problem of rising carbon dioxide content of the atmosphere. The conferees were ecologists, chemists, physicists, and others with particular experience and interest in the problem....

It is known that the carbon dioxide situation, as it has been observed within the last century, is one which might have considerable biological, geographical, and economic consequences within the not too distant future. What is important is that with the rise of carbon dioxide, by way of exhaust gases from engines and other sources, there is a rise in the temperature of the atmosphere and oceans. It is estimated that a doubling of the carbon dioxide content of the atmosphere would produce an average atmosphere temperature rise of 3.8 degrees Fahrenheit. This could be enough to bring about an immense flooding of the lower portions of the world's land surface, resulting from increased melting of glaciers. So far, the increase of carbon dioxide has been of the order of 10 percent, and the oceans are already experiencing some rise of temperature.

A principal purpose of the meeting was to discuss the subject with a view to clarifying the minds of a small interdisciplinary group and crystallizing some ideas for future scientific research. It is hoped that the publication of this summary of conference discussions may contribute to further examination of the carbon dioxide situation. The subject should be one of considerable concern and controversy.

INTRODUCTION AND SUMMARY

Carbon dioxide is not a pollutant in the ordinary sense. It is colorless and odorless. It has no immediate nasty effects. Even the largest amount likely to accumulate in the atmosphere, if the entire reserve of fossil fuels were burned, would not be detrimental to the existence of life; in fact, plant life would be more luxuriant. It is an inevitable product of combustion and cannot be filtered out or precipitated out. Ordinary pollutants are washed out of the atmosphere after a month or so; carbon dioxide will continue to accumulate as long as fossil fuels continue to be burned at present rates.

There is a lack of exact knowledge of the carbon cycle which is part of the general lack of quantitative knowledge of the biogeochemistry of the earth. The increasing funds available for general research and the improving coordination of research effort should help to reduce the uncertainties about the implications of the rise in atmospheric carbon dioxide.

It seems quite certain that a continuing rise in the amount of atmospheric carbon dioxide is likely to be accompanied by a significant warming of the surface of the earth which by melting the polar ice caps would raise sea level and by warming the oceans would change considerably the distributions of marine species including commercial fisheries.

The biogeochemical system of the surface of the earth is, in general, very stable and has persisted with little change over geologically long periods of time. However, the buffering mechanisms which have been adequate in the past seem unlikely to be able to compensate fully for the changes of the magnitude of those now being effected by man.

The effects of a rise in atmospheric carbon dioxide are world-wide. They are significant not to us but to the generations to follow. The consumption of fossil fuel has increased to such a pitch within the last half century that the total atmospheric consequences are matters of concern for the planet as a whole. Although there is the possibility of capturing the CO_2 formed by the burning of fossil fuels and storing it in the form of carbonates, relief is most likely through the development of new sources of power....

PROBLEMS AND NEEDS

The most alarming thing about the increase of CO_2 is how little is actually known about it. Part of this lack is the general absence of detailed knowledge about the surface of the earth and changes taking place there. We should know what is happening to the physical environment both as a result of man's activity and of natural causes. There has been very little consideration of the biological consequences of man's manipulation of the environment or of the biological changes which result from climatic change. For example, the distributions of most marine organisms, including many species of commercial importance, are not known. There is evidence that the warming in the northern hemisphere during the twentieth century has drastically changed the ranges of many fish, but these changes have been little studied or cataloged.

The Continuation of the CO_2 Monitoring Program

It might be argued that there is little immediate danger in a rise in CO_2, that the world can afford to sit back for a few decades to see what happens; measurements of CO_2 could be abandoned for a while because sufficient information could be obtained from a comparison of present values with those 25 or 30 years from now. However, careful measurements have been made for a period of less than ten years in only three places in the world. (CO_2 has been measured in other places for short periods.) This is too short a time to be sure that there is no immediate cause for alarm. (In addition, many things could be learned about the relationship among atmosphere, hydrosphere, and biosphere.) . . .

Science, Technology and Environment

Air pollution in the ordinary sense does not include the CO_2 rise in the atmosphere (although CO_2 can be used as an indicator of pollution). Man-made haze is world-wide in distribution. Cities in the United States have begun to realize that their problem is their neighbor's problem, and the other way 'round, but they still have not seemed to realize that more cars make the haze thicker or that monitoring devices should be between cities and not downtown. To a great many people no smoke still means no work and hard times. Pollution is now a political and social problem far more than it is a scientific one. (In fact, visible air pollution is a valuable tool for the meteorologist interested in studying circulation.)

Fossil fuel residues in the atmosphere in the form of such things as carcinogenic hydro-carbons are becoming more and more serious but these can be removed (or prevented from reaching the open air). CO_2 might be trapped and stored as carbonate but the expense would be prohibitive. It is almost inevitable that as long as we continue to rely heavily on fossil fuels for our increasing power needs, atmospheric CO_2 will continue to rise and the earth will be changed, more than likely for the worse.

In terms of the health of the planet there are no underdeveloped countries, only an increasing number of overdeveloped ones. Environmental damage is being done by modern technology at an increasing rate with no predictions of environmental consequences. Artificial photosynthesis has been proposed as a solution to the world's food problem. This could lead to a large reduction in "unnecessary" plant cover and an accelerated rise in atmospheric CO_2.

The fact that there may be a problem is beginning to be understood. Technical conferences, such as those held this year by the American Society of Civil Engineers and the New York Academy of Medicine, are considering environmental effects, but still only the short range effects.

The potentially dangerous increase of CO_2, due to the burning of fossil fuels, is only one example of the failure to consider the consequences of industrialization and economic development. Mankind is not helped by any technique which leads to short range benefits but long range dangers. Overdevelopment and concomitant overpopulation in many areas of the world have made the problem not one of increasing productivity but one of preventing further decreases.

Man's ability to change the environment has increased greatly in the last sixty years and is likely to continue to increase for some time to come. Even now it is almost impossible to predict all of the consequences of man's activities. It is possible, however, to predict that there will be problems without being specific about it. It is very important to alert more people, more scientists and more scholars in the social sciences as well as the pragmatical sciences, to the need for planning and the realization that there is an obligation to provide for the future as well as the present.

PRESIDENT'S SCIENCE ADVISORY COMMITTEE

RESTORING THE QUALITY OF OUR ENVIRONMENT

By the middle of the 1960s, the health of the natural environment had emerged as a major mainstream concern in American politics. Rachel Carson's 1962 *Silent Spring* famously alerted the nation to the ills of pesticides and other toxins as they flowed through natural and human ecosystems, and the word *ecology* became a critical buzzword among critics of Americans' relationships with their natural surroundings.* At the fringes of this mainstream conversation about ecology sat climate scientists like Roger Revelle and Charles Keeling, who began to see CO_2 and climate change as an important part of the same conversation. When, in 1965, the Johnson administration commissioned a report on the problems and potential for America's natural environment, the president tapped Revelle, Keeling, geochemist Harmon Craig, oceanographer Wallace Broecker, and renowned climate modeler Joseph Smagorinsky to explain where CO_2 and climate change fit into the pollution picture. To what extent did their understanding of the problem—and the specific scientific recommendations they made for addressing it—reflect their scientific sensibilities, and to what extent did their framing of the issue reveal their concern over the environmental impacts of CO_2 and climate change? Were they acting as scientists, environmentalists, concerned citizens, or a mix of the three? The excerpt here begins with the president's introductory letter and the introduction

---

President's Science Advisory Committee, *Restoring the Quality of Our Environment: Report of the Pollution Panel* (1965). Excerpts: President's letter, pp. x, 1–2, 112–13, 123–24, 126–27.

* Rachel Carson, *Silent Spring* (Boston: Houghton Mifflin, 1962).

to the larger report, penned by the chair of Johnson's Science Advisory Committee, Donald Hornig. It also includes the recommendations of the carbon dioxide panel, identified in the "list of participants" in the prefatory material.

• •

The White House
Washington
November 5, 1965

Ours is a nation of affluence. But the technology that has permitted our affluence spews out vast quantities of wastes and spent products that pollute our air, poison our waters, and even impair our ability to feed ourselves. At the same time, we have crowded together into dense metropolitan areas where concentration of wastes intensifies the problem.

Pollution now is one of the most pervasive problems of our society. With our numbers increasing, and with our increasing urbanization and industrialization, the flow of pollutants to our air, soils and waters is increasing. This increase is so rapid that our present efforts in managing pollution are barely enough to stay even, surely not enough to make the improvements that are needed.

Looking ahead to the increasing challenges of pollution as our population grows and our lives become more urbanized and industrialized, we will need increased basic research in a variety of specific areas, including soil pollution and the effects of air pollutants on man. We must give highest priority of all to increasing the numbers and quality of the scientists and engineers working on problems related to the control and management of pollution. . . .

INTRODUCTION

Environmental pollution is the unfavorable alteration of our surroundings, wholly or largely as a byproduct of man's actions, through direct or indirect effects of changes in energy patterns, radiation levels, chemical and physical constitution and abundances of organisms. These changes may affect man directly, or through his supplies of water and of agricultural and other biological products, his physical objects or possessions, or his opportunities for recreation and appreciation of nature.

The production of pollutants and an increasing need for pollution management are an inevitable concomitant of a technological society with a high standard of living. Pollution problems will increase in importance as our technology and standard of living continue to grow.

Our ancestors settled in a fair and unspoiled land, easily capable of absorbing the wastes of its animal and human populations. Nourished by the resources of this continent, the human inhabitants have multiplied greatly and have grouped themselves to form gigantic urban concentrations, in and around which are vast and productive industrial and agricultural establishments, disposed with little regard for state or municipal boundaries.

Huge quantities of diverse and novel materials are dispersed, from city and farm alike, into our air, into our waters and onto our lands. These pollutants are either unwanted by-products of our activities or spent substances which have served intended purposes. By remaining in the environment they impair our economy and the quality of our life. They can be carried long distances by air or water or on articles of commerce, threatening the health, longevity, livelihood, recreation, cleanliness and happiness of citizens who have no direct stake in their production, but cannot escape their influence.

Pollutants have altered on a global scale the carbon dioxide content of the air and the lead concentrations in ocean waters and human populations. Pollutants have reduced the productivity of some of our finest agricultural soils, and have impaired the quality and the safety of crops raised on others. Pollutants have produced massive mortalities of fishes in rivers, lakes and estuaries and have damaged or destroyed commercial shellfish and shrimp fisheries. Pollutants have reduced valuable populations of pollinating and predatory insects, and have appeared in alarming amounts in migratory birds. Pollutants threaten the estuarine breeding grounds of valuable ocean fish; even Antarctic penguins and Arctic snowy owls carry pesticides in their bodies.

The land, water, air and living things of the United States are a heritage of the whole nation. They need to be protected for the benefit of all Americans, both now and in the future. The continued strength and welfare of our nation depend on the quantity and quality of our resources and on the quality of the environment in which our people live.

The pervasive nature of pollution, its disregard of political boundaries including state lines, the national character of the technical, economic and political problems involved, and the recognized Federal responsibilities for administering vast public lands which can be changed by pollution, for carrying out large enterprises which can produce pollutants, for preserving and improving the nation's natural resources, all make it mandatory that the Federal Government assume leadership and exert its influence in pollution abatement on a national scale.

We attempt here to describe the problem, to distinguish between what is known and what is not, and to recommend steps necessary to assure the lessening of pollution already about us and to prevent unacceptable environmental deterioration in the future....

APPENDIX Y4: ATMOSPHERIC CARBON DIOXIDE

Section 1. Carbon Dioxide from Fossil Fuels—the Invisible Pollutant

Only about one two-thousandth of the atmosphere and one ten-thousandth of the ocean are carbon dioxide. Yet to living creatures, these small fractions are of vital importance. Carbon is the basic building block of organic compounds, and land plants obtain all of their carbon from atmospheric carbon dioxide. Marine plants obtain carbon from the dissolved carbon dioxide in sea water, which depends for its concentration on an equilibrium with the carbon dioxide of the atmosphere. Marine and terrestrial animals, including man, procure, either directly or indirectly, the substance of their bodies and the energy for living from the carbon compounds made by plants.

All fuels used by man consist of carbon compounds produced by ancient or modern plants. The energy they contain was originally solar energy, transmuted through the biochemical process called photosynthesis. The carbon in every barrel of oil and every lump of coal, as well as in every block of limestone, was once present in the atmosphere as carbon dioxide....

Throughout most of the half-million years of man's existence on earth, his fuels consisted of wood and other remains of plants which had grown only a few years before they were burned. The effect of this burning on the content of atmospheric carbon dioxide was negligible, because it only slightly speeded up the natural decay processes that continually recycle carbon from the biosphere to the atmosphere. During the last few centuries, however, man has begun to burn the fossil fuels that were locked in the sedimentary rocks over five hundred million years, and this combustion is measurably increasing the atmospheric carbon dioxide.

In the geologic past, the quantity of carbon dioxide in the atmosphere was determined by the equilibrium between rates of weathering and photosynthesis, and the rate of injection of volcanic carbon dioxide.... On a human scale, the times involved are very long. The known amounts of limestone and organic carbon in the sediments indicate that the atmospheric carbon dioxide has been changed forty thousand times during the past four billion years, consequently the residence time of carbon in the atmosphere, relative to sedimentary rocks, must be of the order of a hundred thousand years.... Within a few short centuries, we are returning to the air a significant part of the carbon that was slowly extracted by plants and buried in the sediments during half a billion years....

Other Possible Effects of an Increase in Atmospheric Carbon Dioxide

Melting of the Antarctic ice cap. It has sometimes been suggested that atmospheric warming due to an increase in the CO_2 content of the atmosphere may result in a catastrophically rapid melting of the Antarctic ice cap, with an accompanying rise in sea level. From our knowledge of events at the end of the Wisconsin period, 10 to 11 thousand years ago, we know that melting of continental ice caps can occur very rapidly on a geologic time scale. But such melting must occur relatively slowly on a human scale....

Rise of sea level. The melting of the Antarctic ice cap would raise sea level by 400 feet. If 1,000 years were required to melt the ice cap, the sea level would rise about 4 feet every 10 years, 40 feet per century. This is a hundred times greater than present worldwide rates of sea level change.

Warming of sea water. If the average air temperature rises, the temperature of the surface ocean waters in temperate and tropical regions could be expected to rise by an equal amount. (Water temperatures in the polar regions are roughly stabilized by the melting and freezing of ice.) An oceanic warming of 1° to 2°C (about 2°F) occurred in the North Atlantic from 1880 to 1940. It had a pronounced effect on the distribution of some fisheries, notably the cod fishery, which has greatly increased around Greenland and other far northern waters during the last few decades. The amelioration of oceanic climate also resulted in a marked retreat of sea ice around the edges of the Arctic Ocean.

Increased acidity of fresh waters. Over the range of concentrations found in most soil and ground waters, and in lakes and rivers, the hydrogen ion concentration varies nearly linearly with the concentration of free CO_2. Thus the expected 25% increase in atmospheric CO_2 concentration by the end of this century should result in a 25% increase in the hydrogen ion concentration of natural waters or about a 0.1 drop in pH. This will have no significant effect on most plants.

Increase in photosynthesis. In areas where water and plant nutrients are abundant, and where there is sufficient sunlight, carbon dioxide may be the limiting factor in plant growth. The expected 25% increase by the year 2000 should significantly raise the level of photosynthesis in such areas. Although very few data are available, it is commonly believed that in regions of high plant productivity on land, such as the tropical rain forests, phosphates, nitrates and other plant nutrients limit production rather than atmospheric CO_2. This is probably also true of the oceans. . . .

Conclusions and Findings

Through his worldwide industrial civilization, Man is unwittingly conducting a vast geophysical experiment. Within a few generations he is burning the fossil fuels that slowly accumulated in the earth over the past 500 million years. The CO_2 produced by this combustion is being injected into the atmosphere; about half of it remains there. The estimated recoverable reserves of fossil fuels are sufficient to produce nearly a 200% increase in the carbon dioxide content of the atmosphere.

By the year 2000 the increase in atmospheric CO_2 will be close to 25%. This may be sufficient to produce measurable and perhaps marked changes in climate, and will almost certainly cause significant changes in the temperature and other properties of the stratosphere. At present it is impossible to predict these effects quantitatively, but recent advances in mathematical modeling of the atmosphere, using large computers, may allow useful predictions within the next 2 or 3 years.

Such predictions will need to be checked by careful measurements: a series of precise measurements of the CO_2 content in the atmosphere should continue to be made by the U.S. Weather Bureau and its collaborators, at least for the next several decades; studies of the oceanic and biological processes by which CO_2 is removed from and added to the atmosphere should be broadened and intensified; temperatures at different heights in the stratosphere should be monitored on a worldwide basis.

The climatic changes that may be produced by the increased CO_2 content could be deleterious from the point of view of human beings. The possibilities of deliberately bringing about countervailing climatic changes therefore need to be thoroughly explored. A change in the radiation balance in the opposite direction to that which might result from the increase of atmospheric CO_2 could be produced by raising the albedo, or reflectivity, of the earth. Such a change in albedo could be brought about, for example by spreading very small reflecting particles over large oceanic areas. The particles should be sufficiently buoyant so that they will remain close to the sea surface and they should have a high reflectivity, so that even a partial covering of the surface would be adequate to produce a marked change in the amount of reflected sunlight. Rough estimates indicate that enough particles partially to cover a square mile could be produced for perhaps one hundred dollars. Thus a 1% change in reflectivity might be brought about for about 500 million dollars a year, particularly if the reflecting particles were spread in low latitudes, where the incoming radiation is concentrated. Considering the extraordinary economic and human importance of climate, costs of this magnitude do not seem excessive. An early development of the needed technology might have other uses, for example in inhibiting the formation of hurricanes in tropical oceanic areas.

DONELLA H. MEADOWS, DENNIS L. MEADOWS,
JØRGEN RANDERS, AND WILLIAM W. BEHRENS III

THE LIMITS TO GROWTH

*A Report for the Club of Rome's Project
on the Predicament of Mankind*

Environmentalism and the "ecology movement" gained momentum in the late 1960s and early 1970s, and by 1972 environmental issues had become major agenda items in both domestic and international political discussions. If the domestic buzzword of the 1960s was *pollution*, however, the international discussion of environmental issues in the 1970s revolved around the concept of limits—limits to food supplies, limits to an ecosystem's ability to absorb pollution, limits to population levels, and limits to natural resources. In 1972, a group called the Club of Rome published a report on a series of interrelated social, economic, and environmental problems that they referred to as the "world problematique." Authored by a group of MIT systems scientists, including Donella Meadows, Dennis Meadows, Jørgen Randers, and William Behrens III, *The Limits to Growth* became one of the touchstone works of 1970s environmentalism, selling over ten million copies in twenty-five languages and engendering decades of debate about its methodology and conclusions. Here the authors explain the nature and purpose of the study. How does CO_2 fit into the larger picture of exponential growth and its limits, and why did scientists and environmentalists in the 1970s have

---

Donella H. Meadows, Dennis L. Meadows, Jørgen Randers, and William W. Behrens III, *The Limits to Growth: A Report for the Club of Rome's Project on the Predicament of Mankind* (New York: Signet, 1972). Excerpt: pp. 9–12, 23–24, 71–73.

an easier time framing CO_2 within the discourse of limits than they did within the discourse of domestic pollution?*

• •

FOREWORD BY WILLIAM WATTS, POTOMAC ASSOCIATES

In April 1968, a group of thirty individuals from ten countries—scientists, educators, economists, humanists, industrialists, and national and international civil servants—gathered in the Accademia dei Lincei in Rome. They met at the instigation of Dr. Aurelio Peccei, an Italian industrial manager, economist, and man of vision, to discuss a subject of staggering scope—the present and future predicament of man.

The Club of Rome

Out of this meeting grew The Club of Rome, an informal organization that has been aptly described as an "invisible college." Its purposes are to foster understanding of the varied but interdependent components—economic, political, natural, and social—that make up the global system in which we all live; to bring that new understanding to the attention of policy-makers and the public worldwide; and in this way to promote new policy initiatives and action.

The Club of Rome remains an informal international association, with a membership that has now grown to approximately seventy persons of twenty-five nationalities. None of its members holds public office, nor does the group seek to express any single ideological, political, or national point of view. All are united, however, by their overriding conviction that the major problems facing mankind are of such complexity and are so interrelated that traditional institutions and policies are no longer able to cope with them, nor even to come to grips with their full content. . . .

---

* For more on *The Limits to Growth*, see Fernando Elichirigoity, *Planet Management: Limits to Growth, Computer Simulation, and the Emergence of Global Spaces* (Evanston, IL: Northwestern University Press, 1999).

The Project on the Predicament of Mankind

A series of early meetings of The Club of Rome culminated in the decision to initiate a remarkably ambitious undertaking—the Project on the Predicament of Mankind.

The intent of the project is to examine the complex of problems troubling men of all nations: poverty in the midst of plenty; degradation of the environment; loss of faith in institutions; uncontrolled urban spread; insecurity of employment; alienation of youth; rejection of traditional values; and inflation and other monetary and economic disruptions. These seemingly divergent parts of the "world problematique," as The Club of Rome calls it, have three characteristics in common: they occur to some degree in all societies; they contain technical, social, economic, and political elements; and, most important of all, they interact.

It is the predicament of mankind that man can perceive the problematique, yet, despite his considerable knowledge and skills, he does not understand the origins, significance, and interrelationships of its many components and thus is unable to devise effective responses. This failure occurs in large part because we continue to examine single items in the problematique without understanding that the whole is more than the sum of its parts, that change in one element means change in the others....

The Phase One study was conducted by an international team under the direction of Professor Dennis Meadows, with financial support from the Volkswagen Foundation. The team examined the five basic factors that determine, and therefore, ultimately limit, growth on this planet—population, agricultural production, natural resources, industrial production, and pollution. The research has now been completed. This book is the first account of the findings published for general readership.

INTRODUCTION

... It is not the purpose of this book to give a complete, scientific description of all the data and mathematical equations included in the world model. Such a description can be found in the final technical report of our project. Rather, in *The Limits to Growth* we summarize the main features of the model and our findings in a brief, nontechnical way. The emphasis is meant to be not on the equations or the intricacies of the model, but on what it tells us about the world. We have used a computer as a tool to aid our own understanding of the causes and consequences of the accelerating trends that characterize the modern world, but familiarity with computers is by no means necessary to comprehend or to discuss our conclusions. The implications of those accelerating trends raise issues that go far beyond the proper domain of a purely scientific document. They must be debated by a wider community than that of scientists alone. Our purpose here is to open that debate. . . .

Our conclusions are:

1. If the present growth trends in world population, industrialization, pollution, food production, and resource depletion continue unchanged, the limits to growth on this planet will be reached sometime within the next one hundred years. The most probable result will be a rather sudden and uncontrollable decline in both population and industrial capacity.
2. It is possible to alter these growth trends and to establish a condition of ecological and economic stability that is sustainable far into the future. The state of global equilibrium could be designed so that the basic material needs of each person on earth are satisfied and each person has an equal opportunity to realize his individual human potential.
3. If the world's people decide to strive for this second outcome rather than the first, the sooner they begin working to attain it, the greater will be their chances of success.

These conclusions are so far-reaching and raise so many questions for further study that we are quite frankly overwhelmed by the enormity of the job that must be done. We hope that this book will serve to interest other people, in many fields of study and in many countries of the world, to raise the space and time horizons of their concerns and to join us in understanding and preparing for a period of great transition—the transition from growth to global equilibrium. . . .

THE NATURE OF EXPONENTIAL GROWTH

... Virtually every pollutant that has been measured as a function of time appears to be increasing exponentially. The rates of increase of the various examples shown below vary greatly, but most are growing faster than the population. Some pollutants are obviously directly related to population growth (or agricultural activity, which is related to population growth). Others are more closely related to the growth of industry and advances in technology. Most pollutants in the complicated world system are influenced in some way by both the population and the industrialization positive feedback loops.

Let us begin by looking at the pollutants related to mankind's increasing use of energy. The process of economic development is in effect the process of utilizing more energy to increase the productivity and efficiency of human labor. In fact, one of the best indications of the wealth of a human population is the amount of energy it consumes per person. Per capita energy consumption in the world is increasing at a rate of 1.3 percent per year, which means a total increase, including population growth, of 3.4 percent per year.

At present about 97 percent of mankind's industrial energy production comes from fossil fuels (coal, oil, and natural gas). When these fuels are burned, they release, among other substances, carbon dioxide (CO_2) into the atmosphere. Currently about 20 billion tons of CO_2 are being released from fossil fuel combustion each year.... The measured amount of CO_2 in the atmosphere is increasing exponentially, apparently at a rate of about 0.2 percent per year. Only about one half of the CO_2 released from burning fossil fuels has actually appeared in the atmosphere—the other half has apparently been absorbed, mainly by the surface water of the oceans.

If man's energy needs are someday supplied by nuclear power instead of fossil fuels, this increase in atmospheric CO_2 will eventually cease, one hopes before it has had any measurable ecological or climatological effect.

STUDY OF MAN'S IMPACT ON CLIMATE

INADVERTENT CLIMATE MODIFICATION

Around the same time that the Club of Rome sought to map out the larger "predicament of mankind," atmospheric scientists also began to put together a more integrated picture of the physical systems they studied. Sponsored in large part by MIT—a hub of systems science—the Study of Man's Impact on Climate (SMIC) in Stockholm, Sweden, brought together some of the world's top atmospheric and environmental scientists to articulate the myriad ways in which human activities, including the addition of CO_2 via the burning of fossil fuels, might affect global atmospheric processes. Here, the SMIC group frames the interrelated nature of atmospheric problems in familiar environmental terms, and then articulates the state of the art in an interdisciplinary, systems science understanding of atmospheric processes. Note the study's highly unconventional epigraph. What does the epigraph reveal about the relationships between climate science and environmentalism in the early 1970s? How about the report's recommendations?

. . . .

Carroll L. Wilson and William H. Matthews, eds., *Inadvertent Climate Modification: Report of the Study of Man's Impact on Climate (SMIC): Proceedings of the Conference Held in Stockholm, June 28 to July 16, 1970* (Cambridge, MA: MIT Press, 1971). Excerpt: frontispiece, pp. xv, 11–15. Copyright 1971, Massachusetts Institute of Technology, by permission of the MIT Press.

SAMUDRA MÈKHALÈ DEVI PARVATHA STHANA MANDALÈ
PĀDA SPARSAM KSHMASWA MÈ

Oh, Mother earth, ocean-girdled and mountain-breasted, pardon me for trampling on you.

Sanskrit prayer

PREFACE

The need for the 1971 Study of Man's Impact on Climate (SMIC) was perceived by several of us who had been deeply involved in the planning and conduct of the 1970 Study of Critical Environmental Problems (SCEP). SCEP, which had been sponsored by M.I.T. and had been participated in by more than 70 U.S. scientists, examined global environmental problems in some detail and published the results in October 1970. Following that intensive and extensive exercise, we concluded that an examination of the critical questions associated with the climatic effects of man's activities should be conducted by scientists from many nations at the earliest possible date.

The implications of inadvertent climate modification, both in terms of direct impact on man and the biosphere and of the hard choices that societies might face to prevent such impacts, are profound. Should preventive or remedial action be necessary, it will almost certainly require effective cooperation among the nations of the world. SMIC was developed to assist in this process by providing an international scientific consensus on what we know and do not know and how to fill the gaps. It is hoped that this consensus will provide an important input into planning for the 1972 United Nations Conference on the Human Environment and for numerous other national and international activities.

The committee that planned SMIC during the fall of 1970 and the spring of 1971 was chaired by Professor Carroll L. Wilson, M.I.T., and included Dr. William W. Kellogg, National Center for Atmospheric Research; Dean Thomas Malone, University of Connecticut; Professor William H. Matthews, M.I.T.; and Dr. G. D. Robinson, Center for the Environment and Man, Inc.

The following scientific objectives were adopted for SMIC: to review SCEP findings critically; to point to global environmental problems that were slighted or overlooked; to obtain a more complete assessment of present knowledge; to amplify certain points, especially with respect to establishing priorities among problems and determining how to proceed in implementing recommendations for action and further extensive research; and to point to questions requiring international policy decisions....

2.2 MAN'S ACTIVITIES INFLUENCING CLIMATE

One of the most clearly evident influences of man on his environment is his pollution of the atmosphere. In cities the pollution is sometimes acute, but even in the most remote places in the world it is still possible to detect traces of man-made contaminants in the air. The resulting rise in the particulate load of the atmosphere can be attributed to both industrial activity and the burning of waste crops and vegetation that is a practice in many tropical areas. Particles scatter and absorb solar radiation and also have an effect on the outgoing infrared radiation from the surface, so these man-made products will influence the heat balance over wide areas.

Some gaseous constituents of the atmosphere also absorb solar and infrared radiation, and carbon dioxide (CO_2), water vapor (H_2O), and ozone (O_3) are in this category. It is well known that the CO_2 content of the global atmosphere has been rising due to the burning of fossil fuels—coal, petroleum, and natural gas—and it is expected that it will go up about another 20 percent by 2000 A.D.

Until quite recently it has been assumed that man could not compete directly with nature in the release of heat on a large scale, but now we must take a further look at this matter as we realize the implications of a doubling of the present world population of 3.6 billion by the year 2000, coupled with an expectation of more energy to be used per capita. The production of energy of all sorts is rising at a rate of from 5 to 6 percent per year for the world (5.5 percent per year is equivalent to a factor of 5 in 30 years, or by the year 2000 A.D.). There may eventually be industrialized areas of 10^3 to 10^5 km^2 where the additional input of energy by man will be equivalent to the net radiation from the sun; and on a continental scale the present insignificant contribution may rise to 1 percent of the continental net radiation average after about 40 years.

We have spoken of some of the effects of industrial activity, but for thousands of years before the Industrial Revolution agricultural and animal grazing practices have had a profound influence on large regions of the world, and it seems very likely that these have already resulted in changes of climate of those regions. Grazing by goats and other domestic animals has reduced parts of Africa and Southwest Asia to semideserts; dense forests of the mountainous areas from Turkey to Afghanistan and of the Mediterranean, Europe, and the eastern United States have been cut down to make arable or grazing land; and the savannah grasslands of the tropics are nearly all man-made. The net result is that some 20 percent of the total area of the continents has been drastically changed, with a consequent change in the heat and water budget. In arid or semiarid areas the growing demands for water for irrigation have reduced the reserves of groundwater, and the process of irrigation increases the water vapor in the air.

Another influence of man is his manipulation of the surface waters by building dams, creating lakes, draining swamps, and diverting rivers. Artificial lakes, for example, will alter the heat balance of the area because water has a lower albedo; it has a much greater heat capacity, and it adds water vapor to the air.

Perhaps the diversion of rivers from one region to another has even greater potential implications, since the water so diverted will convert substantial dry desert or semidesert dry areas to irrigated farmland, and three-fourths to nine-tenths of irrigation water is evaporated in the air.

Control of river discharge into ocean areas subject to winter freezing could greatly influence the rate of freezing or melting. Such activities coupled with intentional dusting of sea ice to hasten melting could have serious regional and even global repercussions.

Modification of precipitation by seeding clouds with freezing nuclei has been sufficiently attractive to encourage the practice throughout the world. Any change in precipitation pattern influences the heat budget of the atmosphere; therefore, widespread seeding operations, including efforts to change the course of hurricanes, may modify the normal patterns of rainfall and snowfall enough to have an influence on the heat budget of a part of the atmosphere.

Finally we come to one of the most rapidly escalating activities of man, his various modes of transportation. Automobiles contribute approximately half of some air pollutants observed in U.S. cities, and the same is true in most other industrialized countries. Since the automobile exhaust forms "smog" particles, this is a major contribution to the particle increase that we have already discussed. It is also interesting to note that roads in the United States cover nearly 1 percent of the entire area. This is not as big a change of land use as that due to agriculture, which has been discussed earlier, but it is not entirely trivial.

Aircraft also produce exhaust products, and it is currently estimated that commercial aviation will double its jet fuel consumption every five or six years in the next decade or so. These exhaust products from jet airliners are deposited high in the troposphere or the lower stratosphere. There are indications that jet traffic has already caused a small increase in cirrus cloudiness in heavily traveled areas, and this will have a small effect on the heat balance of the atmosphere. The supersonic transports fly in a region where the average residence time of their exhaust products is one or two years, so there is a chance for the concentration to build up.

In speaking about the future of man and his climate we are faced with two important classes of consideration: meteorological (or geophysical) and social. Whereas we do have a rudimentary theory of climate and a good deal of knowledge about how things behave in the physical world, when we turn to a forecast of man's behavior we must resort to simple extrapolation of present trends with only slight shadings to take account of the more obvious interactions that we can foresee. There are no laws that we know of, or mathematical models, that will allow us to predict the future course of human affairs better than such extrapolations. Therefore, we can in the end only forecast what *could* happen *if* mankind proceeds to act in a certain way, more or less as he is acting now.

2.3 PRESENT THEORY AND MODELS OF CLIMATE CHANGE

2.3.1 *Discussion*

Climate is determined by a balance among numerous interacting physical processes in the oceans and atmosphere and at the land surface. Locally and globally, climate is subject to change on all scales of time, but the physical processes themselves remain the same and are amenable to study by statistical, physical, and mathematical techniques. If we are to assess the possibility and nature of a man-made climatic change, we must understand how the physical processes produce the present climate and also how past changes of climate, clearly not man-made, have occurred.

Several physical-mathematical techniques—they have come to be called "models"—are being developed to attack the problem. We have distinguished four types: (1) "global-average" models, in which horizontal motion of the atmosphere is neglected; (2) parameterized semiempirical models that consider the whole atmosphere and surface but simulate some of the effects of atmosphere and oceanic motions with the aid of empirically adjusted constants; (3) statistical-dynamical models in which physical laws are applied to statistics of the atmospheric variables; and (4) explicit numerical models in which motions and interactions are treated in detail by integrating mathematical equations expressing the time rate of change of the variables, though in practice the detailed treatment must be abandoned at some minimum scale of motion and replaced by empirical parameterization or statistical methods. . . .

We now know enough of the theory of climate and the construction of climatic models to recognize the possibility of man-made climatic change and to have some confidence in our ability ultimately to compute its magnitude. Recent results obtained using empirical models have in our view increased the urgency of the study of climate theory in its own right. Some of these results, for example, concern the delicate balance of the processes that maintain the arctic sea ice. The empirical models suggest that a small change in mean air temperature or in solar radiation reaching the surface could result in considerable expansion or contraction of the ice pack—a climatic change of great significance to human life. This in our opinion reinforces the urgency of studies of climate by all available methods in order to understand its natural and possible man-made changes.

2.3.2 Recommendations

Data must be gathered judiciously for the understanding of climate change and for the verification of the accuracy of atmosphere-ocean climate models. We therefore recommend:

1. Monitoring the temporal and geographical distribution of the earth-atmosphere albedo and outgoing flux over the entire globe, with an accuracy of at least 1 percent.
2. Monitoring with high resolution the global distribution (horizontal and vertical) of cloudiness, and the extent of polar ice and snow cover with less resolution.
3. Measuring the distribution, optical properties, and trends of atmospheric particles and clouds over the globe.

THE SIERRA CLUB

INTERNATIONAL COMMITTEE QUESTIONNAIRE—FIVE-YEAR PLAN

As the next three documents show, domestic environmental organizations like the Sierra Club worked hard in the 1970s to incorporate global environmental issues into their increasingly international agendas. In 1976, for example, the Sierra Club's International Committee prepared a "Five-Year Plan" to structure the club's future involvement in international environmental affairs. Program director Patricia Scharlin's report on a questionnaire sent to club chapters reveals significant interest in "global commons" issues, but the surveys also reveal some ambivalence about international campaigns among club members. The grassroots structure of the club—an organization made up of local chapters whose financial contributions the club as a whole relied on for operations—made international campaigns a particularly fraught category of action compared to activities like wilderness protection or legal action against domestic polluters. Scharlin's own thinking on the issue, captured in a Venn diagram included in the report, puts the issue of climate change at the center of international environmental protection. Five years later, however, a memo from Sierra Club executive director Michael McCloskey provides some ideas as to why Scharlin's ideas failed to gain traction with the broader club membership. What does McCloskey's memo reveal about the Sierra Club's priorities in the early

---

"Sierra Club International Committee Questionnaire—Five Year Plan," n.d., Five Year Plan, 1976–77, Operational Records, Sierra Club International Program Records, Box 3, Folder 10, Sierra Club Archives.

1980s? Where do you think the International Program developed in the late 1970s fit into those priorities?

• •

The Sierra Club International Committee, in response to a Board of Directors' request, is preparing by next winter (1976) a so-called "Five-Year Plan" to structure and guide further developments of the Club's global conservation efforts. In the process of formulating this plan, the International Committee sent out a questionnaire to various chapter and group officers and other active members of the Club. This paper is an analysis of the responses to that questionnaire.

A total of 54 responses were received. One of those responses was a resume of 25 responses within one group. Several responses represented the consensus of a particular group or chapter executive committee. Some of the responses were made by chapter or group international chairpersons. The remainder were simply responses of interested parties, mostly Club leaders who happened to receive a copy of the questionnaire and took the time to answer it.

First, such a response spectrum could hardly be called a representative sample of the Club's membership. It does expose a large range of responses that have definite value to the International Committee and the Board of Directors who will soon make decisions on the International Committee proposed Five-Year Plan. Second, as in all questionnaire analyses, certain latitude must inevitably be pursued by the questionnaire analyzer, so that written responses can be categorized into some meaningful format. In the process the analyzer tends to introduce some of his own biases. I take full responsibility here, but I have tried to let the respondents' unadulterated answers come through with as little manipulation as possible.

Question 1: "List in the order of importance the international environmental issues you wish the club should tackle. Why is each important?"

Respondents tended to make three separate interpretations of the concept "international environmental issue." These three interpretations are discussed in depth in Appendix A attached. Briefly, group one thought in terms of listing under Priority One some overriding issue which seemed to the respondents in that group to be the basic or root cause of environmental problems. They invariably (the Group One respondents) chose population control as the overriding issue. For example, one respondent in this group wrote: "I am particularly concerned over global population. This seems to me a basic issue which must be solved or all will be lost!" (26).

Group Two tended to distinguish international issues from local issues by the criterion of what kind of government agency might be expected to deal with the particular problem. Thus, international issues were those which could only be dealt with by super-national entities, such as the United Nations, regional organizations, treaty conferences and the like. Ocean and air pollution and preservation were matters of paramount concern for Group Two respondents.

Group Three tended to think of an international environmental issue as a more general category of a local issue. This third interpretation was particularly obvious in respondents' answers who related question 1 to question 2 which concerned local issues of an international nature which their particular chapter or group was working on. These respondents stated that the answer to question 2 was the same as that of question 1.

Table I presents a summary of the answers to question 1 of the questionnaire. Placing Table I in total perspective, 40% of all respondents considered population control as the number one global issue. This probably reflected more than anything the notions the respondents had that population control was the root cause of environmental problems. Thirty percent of all respondents considered protection of our "global commons" as the number one problem priority international issue. This sentiment is probably representative of Group Two thinking, that is, that international environmental issues must be dealt with on a global scale, that is, they are issues that transcend national governments, environmental organizations, and institutions. One should note that 68 responses in all priorities were related to global ocean and air protection. Thus, if one looks at total responses rather than simply first priority responses, then global ocean and air protection becomes the major international environmental issue.

I believe most respondents were very much aware of the interrelated-

TABLE I

QUESTION 1A -- RANK IN PRIORITY OF IMPORTANCE THE 6 INTERNATIONAL ENVIRONMENTAL ISSUES YOU WISH THE CLUB SHOULD TACKLE

General Category / Specific Issue	1	2	PRIORITY 3	4	5	6	Total
Number of Responses on the Questionnaire							
1. Conservation Ethic	2	1	0	2	1	1	7
2. Politics, Economics and the Environment	3	3	3	3	2	4	18
3. Population Control -- Consumption Control							
a. Population Control	21	4	4	1	1	0	31
b. Energy and resource conservation	1	2	1	0	0	2	4
	22	6	5	1	1	2	37
4. The Global Commons -- Ocean and Air							
a. Ocean degradation and pollution	6	7	5	3	2	0	23
b. Protection of marine environment	6	3	3	1	2	0	15
c. Law of the seas	4	2	2	1	1	0	10
d. Air pollution and climatological studies	0	4	3	5	4	2	18
e. Antarctica	0	0	1	1	0	0	2
	16	16	14	11	9	2	68
5. Habitat, Agricultural and Urban Land Use -- The Loss of Land to the Urbanization Process							
a. Wilderness, wetlands, wild rivers, forests, and national parks	2	4	6	5	7	3	27
b. Farmland, soil and water conservation and food production	2	5	3	5	1	4	20
c. Urban land use	0	1	0	0	1	3	5
	4	10	9	10	9	10	52
6. Technology Development and Transfer							
a. Undesirable technologies	1	6	3	3	4	3	20
b. Energy, alternatives, recycling and renewable resources	2	3	8	4	1	2	20
c. Control of existing technologies	1	1	0	5	2	2	11
	4	10	11	12	7	7	51
7. Wildlife and Endangered Species	3	3	5	3	5	2	21
	54	49	47	42	34	28	264

Sierra Club members identified a variety of priorities for international campaigns in the late 1970s, but "air pollution and climatological studies" didn't make the top of anybody's list.

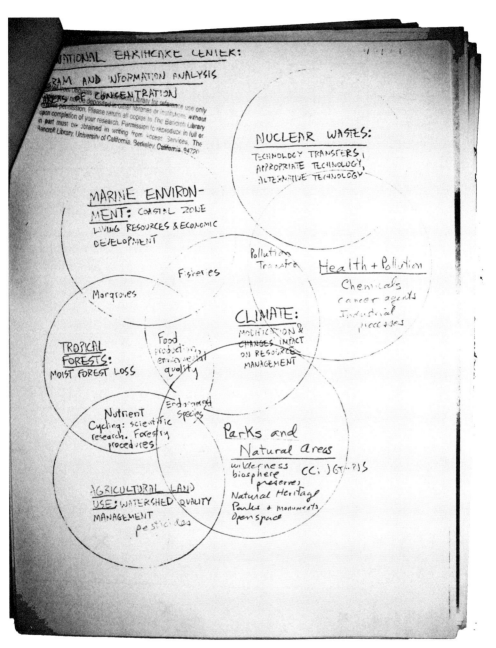

Patricia Scharlin's Venn diagram put climate change at the center of international environmental protection.

ness of environmental issues regardless of their priorities. Thus population growth and growth in resources consumption stimulate political problems and create technologies that impinge upon ocean, air, and land ecosystems. In this sense, priority one issues should not be given undue weight over priority six issues, as it is the whole pattern of issues taken as a totality that is of prime importance. . . .

CATEGORY #4: THE GLOBAL COMMONS—OCEANS AND AIR

This was obviously "the issue" that Sierra Club respondents perceived as belonging expressly to the International Committee. Twenty-six percent of all responses in every priority group related to ocean and air as global problems. In addition, when respondents were asked to "identify international issues of concern in your home area," they picked ocean and air issues over the next most popular category by a margin of 37 to 27.

The breakdown of categories within Category #4 is quite arbitrary. Some of the specific comments were (4a): "ocean degradation," "destruction of oceans, "death of oceans," "ocean pollution," (4b): "conservation of ocean resources—the last food source," "oceans—maximum sustainable yield," "marine environment and pollution," "off-shore fish conservation," "sea-bed and fisheries—international management," "over-fishing and whaling," (4c): "law of the sea," "super-tanker regulation," (4d): "freon-ozone shield," "atmospheric pollution," "clean air," "global atmospheric monitoring," (4e): "Antarctica."

MICHAEL McCLOSKEY

CRITERIA FOR INTERNATIONAL CAMPAIGNS

Michael McCloskey
December 1, 1982

CRITERIA FOR INTERNATIONAL CAMPAIGNS

1. *Goal Is Achievable*—Implications:
   High profile issue of fundamental importance to world community which can be addressed under the following circumstances:
   a. Goal deals with *public policy*; does not deal with private behavior, nor with ultimate physical conditions.
   b. Goals is *clear* and *discrete*; not ill-defined, nor covers impossibly large situations;
   c. Goal deals with conditions for which a legal or regulatory *solution* is *possible*;
   d. An institutional *forum exists* to provide a solution or can be brought into existence; its *powers* are *adequate* to the task or at least can be useful;
   e. *We can* play a pivotal role in *influencing* decisions by that *forum*; our resources are commensurate with task; does not duplicate action by others;

Michael McCloskey, "Criteria for International Campaigns," December 1, 1982, International Committee, 1972–1983, Meetings and Conferences, Operational Files, Sierra Club International Program Records, Box 3, Folder 19, Sierra Club Archives.

f. We are *competent* to undertake the task; we have needed skills or background or can get them; our constituency will feel good about the issues;
g. Issue is *ripe* or can be ripened in reasonable time; there is a convergence of sharpening of the issue, rising public interest, and political will to tackle issue.
2. *Issue can excite Club constituency, especially in U.S.*—implications:
 a. Relates positively to *deeply held values*;
 b. A clear *threat* exists to those values; mere potential problem is insufficient;
 c. *Source* of threat can be *pinpointed*; not populace in general nor generalized conditions;
 d. Problem can be put into a context that is within our constituencies' *realm of experience*; can't be too exotic nor overly technical;
 e. Issue can be dramatized with photographs that symbolize what is at stake with really memorable *images*;
 f. Issue can be updated with news reports or episodes to *build interest in its progress*;
3. *Practical roles exist for U.S. members* (eventually for Sierra Club members in Canada, too)—implications:
 a. U.S. Government is a *major* contributor to problem; or
 b. U.S. Government is *important player* in decisions before forum involved; or
 c. U.S. Government's role is *not a hopelessly overpowered* one;
 d. Procedural *action handles* can be found for our U.S. constituency; e.g., writing to Senators, to President, urging House oversight hearings, testifying, letters-to-editors, boycotts, etc. . . .

NATIONAL CLIMATE PROGRAM ACT OF 1978

Between 1963 and 1985, a number of scientists and environmentalists worked to make climate change a recognizable "environmental" issue. But in the 1970s, interest in climate change had also begun to come from experts outside of scientific and environmental organizations. By 1978, concerns over international droughts and food security, rainmaking and weather modification, the environmental impacts of CO_2-induced climatic change, and the availability of good forecasting information to meet domestic agricultural needs prompted Congress to pass the National Climate Program Act of 1978. Many climate scientists welcomed the act insofar as it promised to streamline climate information and highlight its importance, but the act also reflected a particularly utilitarian view of climate science coming from an agricultural constituency largely divorced from the environmental movement. The act coincided with the beginning of a series of 1979–80 collaborations between the Department of Energy and the American Association for the Advancement of Science, but the constituencies served by the 1978 law and the series of workshops at the end of the decade were very different. How does the NCPA's treatment of climate and climatic change differ from the way scientists like Keeling and environmentalists like Scharlin and McCloskey talk about climate change? What elements do the different framings share in common?

· ·

National Climate Program Act of 1978, Pub. L. No. 95-367, 92 Stat. 601 (1978). Excerpt: pp. 601, 602, 603, 605.

THE NATIONAL CLIMATE PROGRAM ACT OF 1978
PUBLIC LAW 95-367—SEPTEMBER 17, 1978
95TH CONGRESS

An Act

To establish a comprehensive and coordinated national climate policy and program, and for other purposes.

Be it enacted by the Senate and House of Representatives of the United States of America in Congress assembled, That this Act may be cited as the "National Climate Program Act."

SEC. 2. FINDINGS.

The Congress finds and declares the following:

(1) Weather and climate change affect food production, energy use, land use, water resources and other factors vital to national security and human welfare.

(2) An ability to anticipate natural and man-induced changes in climate would contribute to the soundness of policy decisions in the public and private sectors.

(3) Significant improvements in the ability to forecast climate on an intermediate and long-term basis are possible.

(4) Information regarding climate is not being fully disseminated or used, and Federal efforts have given insufficient attention to assessing and applying this information.

(5) Climate fluctuation and change occur on a global basis, and deficiencies exist in the system for monitoring global climate changes. International cooperation for the purpose of sharing the benefits and costs of a global effort to understand climate is essential.

(6) The United States lacks a well-defined and coordinated program in climate-related research, monitoring, assessment of effects, and information utilization.

SEC. 3. PURPOSE.

It is the purpose of the Congress in this Act to establish a national climate program that will assist the Nation and the world to understand and respond to natural and man-induced climate processes and their implications....

SEC. 5. NATIONAL CLIMATE PROGRAM.

... (c) 2 CLIMATE PROGRAM OFFICE.—The Secretary shall establish within the Department of Commerce a National Climate Program Office not later than 30 days after the date of the enactment of this Act. The Office shall be the lead entity responsible for administering the Program. Each Federal officer, employee, department and agency involved in the Program shall cooperate with the Secretary in carrying out the provisions of this Act.

(d) PROGRAM ELEMENTS.—The Program shall include, but not be limited to, the following elements:

(1) assessments of the effect of climate on the natural environment, agricultural production, energy supply and demand, land and water resources, transportation, human health and national security. Such assessments shall be conducted to the maximum extent possible by those Federal agencies having national programs in food, fiber, raw materials, energy, transportation, land and water management, and other such responsibilities, in accordance with existing laws and regulations. Where appropriate such assessments may include recommendations for action;

(2) basic and applied research to improve the understanding of climate processes, natural and man induced, and the social, economic, and political implications of climate change;

(3) methods for improving climate forecasts on a monthly, seasonal, yearly, and longer basis;

(4) global data collection, and monitoring and analysis activities to provide reliable, useful and readily available information on a continuing basis;

(5) systems for the management and active dissemination of climatological data, information and assessments, including mechanisms for consultation with current and potential users;

(6) measures for increasing international cooperation in climate research, monitoring, analysis and data dissemination;

(7) mechanisms for intergovernmental climate-related studies and services including participation by universities, the private sector and others concerned with applied research and advisory services;

(8) experimental climate forecast centers. Which shall (A) be responsible for making and routinely updating experimental climate forecasts of a monthly, seasonal, annual, and longer nature, based on a variety of experimental techniques; (B) establish procedures to have forecasts reviewed and their accuracy evaluated; and (C) protect against premature reliance on such experimental forecasts; and

(9) a preliminary 5-year plan, to be submitted to the Congress for review and comment, not later than 180 days after the enactment of this Act, and a final 5-year plan to be submitted to the Congress not later than 1 year after the enactment of this Act, that shall be revised and extended biennially. Each plan shall establish the goals and priorities for the Program, including the intergovernmental program under section 6, over the subsequent 5-year period, and shall contain details regarding (A) the role of Federal agencies in the programs, (B) Federal funding required to enable the Program to achieve such goals, and (C) Program accomplishments that must be achieved to ensure that Program goals are met within the time frame established by the plan. . . .

(2) The Secretary and the Secretary of State shall cooperate in (A) providing representation at climate-related international meetings and conferences in which the United States participates, and (B) coordinating the activities of the Program with the climate programs of other nations and international agencies and organizations, including the World Meteorological Organization, the International Council of Scientific Unions, the United Nations Environmental Program, the United Nations Educational, Scientific, and Cultural Organization, the World Health Organization, and Food and Agriculture Organization. . . .

(a) GENERAL AUTHORIZATION OF APPROPRIATION.—In addition to any other funds otherwise authorized to be appropriated for the purpose of conducting climate-related programs, there are authorized to be appropriated to the Secretary, for the purpose of carrying out the provisions of this Act, not to exceed $50,000,000 for the fiscal year ending September 30, 1979, and not to exceed $65,000,000 for the fiscal year ending September 30, 1980.

AMERICAN ASSOCIATION FOR
THE ADVANCEMENT OF SCIENCE

ADVISORY GROUP ON CLIMATE MEETING, MAY 26, 1978

In the mid-1970s, the largest and most inclusive of America's science advocacy organizations, the American Association for the Advancement of Science (AAAS), included climate change as a potential subject of focus as it began to consider new directions for the coming decade. In 1978, the AAAS convened an Advisory Group on Climate. The committee comprised researchers from a variety of disciplines associated with climate questions, including, among others, Roger Revelle, leading agronomist and later critic of consensus views on global warming Sylvan Wittwer, Robert White of the National Academy of Sciences Climate Research Board, *Science* editor-in-chief Philip Abelson, and California congressman Harrison Brown's science adviser, Tom Moss. Here, the transcript of the advisory group's first meeting has been digested by an AAAS secretary based on the edited meeting minutes. Why, according to the participants, should a scientific organization like the AAAS care so much about climate change?

· ·

1. The first topic addressed was the question posed by the AAAS Committee on Future Directions: Is climate an appropriate "issue of global concern" to serve as the focus for AAAS involvement with international science?

Partial Transcript of AAAS Advisory Group on Climate meeting, May 26, 1978, AAAS Climate Programs [Box 4], AAAS Archives.

The response of the Advisory Group was a unanimous "yes." Climate was seen as a "very appropriate topic" because it *touches many disciplines*—the physical, biological, and social sciences—and thus *can command the attention of virtually all segments of the AAAS membership*. It involves complex issues of science and public policy, and is of interest to a non-scientific audience, and responds to the "public understanding" facet of AAAS objectives.

Climate was seen as of *great importance*, with wide-ranging and often decisive impacts on human affairs. It is not necessary to suppose that climate is changing or that it can be predicted. The impacts of "normal" variability are far-reaching, being the principal constraint on agricultural productivity and dependability, and with strong effects on the availability and management of resources, such as energy, water, and land, as well as imperfectly understood economic, social, and health and environmental effects. "Climate is not a trivial issue."

Climate was also seen as *fundamentally international*. "It would be hard to find a problem in science that had such inherent global characteristics." Data collection and impact analysis particularly require significant input from all countries and areas, and from virtually all disciplines, and from representatives of the wide range of those affected. . . .

Public implications. The issues of the climate "problem" go far beyond the research of the scientist, and touch international public policy of great importance and sensitivity. "In climate, we need an organization to move into that interface between scientific enquiry and public discussion of policy. This is an area, if you're looking at what is needed internationally, where AAAS could very well fill a gap—one foot in science and one foot into the public understanding area. There *is* a gap and there is a role to be filled, and I think AAAS, if it wished, could usefully fill that role." . . .

2. *An effective international effort requires careful preparation.* "I don't think the federal government is going to be very effective in catalyzing a worldwide scientific program—either data-gathering or data analysis, or impact analysis. A network of non-governmental scientific organizations could be much more effective in organizing worldwide

scientific programs and in building a scientific/political consensus. And worldwide consensus *will* be required. It will do no good for the U.S. scientific community to try to persuade decision-makers of the necessity of change; without a world science consensus on the data and its implications, nothing is going to happen. These are difficult political challenges. The decision to regulate land use, or to store water, or food, or provide money, to tide people over periods of climate stress, will be made only by the 'forcing' function of knowledge. These kinds of actions will be more difficult for politicians than almost anything they are ever asked to do by the scientific community. These difficult decisions will *not* be made unless there is a good scientific reason for the action.

"An important role for AAS would be as a clearing-house for the discipline-oriented societies to interact with the broader aspects of the climate program. But it won't be easy to bridge the jargon-data gap between diverse groups." "We will need to determine what is being planned internationally in the climate area over the next few years, both governmentally and non-governmentally. Then we could determine in a practical way, 'what's missing?'" "Foreign countries always want to know 'what is the U.S. position?' It might be necessary to organize domestically; a U.S. program might be an essential preliminary step before the international catalytic function can work. The international effort might be the *second* step." "AAAS may need to work in close association with an international agency, such as UNEP. A meeting held in conjunction with a wider international gathering might be an economical and effective way to serve the catalytic role. It would also give the climate discussion a broader base of participation." "My experience of UN conferences is that they do tend to be 'set,' with government positions and plenary sessions. The delegates would, I think, welcome an opportunity to attend a lively discussion of a critical world issue." . . .

"It would be important that the disciplines feel AAAS is helping, rather than interfering. If we could demonstrate that farmers are very concerned about, say, run-off, or effective rainfall, or humidity, or variations in solar insolation—then this kind of feed-back would be useful to climatologists. Climatology is a peculiarly applied science in that its raison d'être is its relations to human welfare. You have to find out those things about climate that mean something to somebody." . . .

"Although what is known about climate is developed by scientists, all the decisions based on this knowledge seem to be made by non-specialists—whether in Congress or by the general public. Scientists need to learn from each other about the implications of climate, but it's at least as important for them to learn also from decision-makers—to learn what the real problems are. . . . "

"Restrictions on behavior, or heavy capital investment may be required. Therefore, there will have to be an unusually high degree of unanimity and clarity on the part of the scientific community to obtain from political leaders the decisions and actions required. We'd like to avoid the kinds of controversy and doubt re the validity of the data and the credibility of the scientists that have for years beset the clean air standards." "If the scientific community presents to the public a forecast that claims to have a skill attached to it, but the basis of the skill is not obvious, many will not trust the prediction, and may not use it. There's a *confidence* problem here that goes beyond the basic problem of where you can *produce* such a prediction." "Isn't this question of public confidence a more suitable area for AAAS than the science of climate *per se*? AAAS might also help advise how climatic information could be used more effectively by the government, as well as by all the other users of climatic information, and by those most affected." "But climatic information can be very sensitive—in some cases affecting the political stability of a country. We've see food-importing countries decline to provide climatic data so the food-exporting countries won't raise prices."

DAVID SLADE

ACTION FLOW

U.S. Carbon Dioxide Research and Assessment Program

In 1978, the recommendations of the AAAS Advisory Group on Climate Change bore fruit in the form of a joint project between the AAAS and the Department of Energy. Led on the DoE side by David Slade of the agency's Office of Health and Environmental Research, the program sought to explore the implications of rising atmospheric CO_2 from a wide variety of disciplinary perspectives with the ultimate objective of informing the Carter administration's energy policy. The Department of Energy was a new cabinet-level agency, created in 1977 in an effort to rationalize and consolidate a diverse set of national energy interests from renewable energy research to the management of the U.S. nuclear arsenal. At the end of the first phase of the project, Slade produced this flowchart, a somewhat convoluted schematic for incorporating research on CO_2 into DoE policy under the second Carter administration. What exactly would this sketch of a plan do for climate change policy in the 1980s? How does the chart reflect Slade's objectives as a liaison between the AAAS and DoE? What kind of institutional constraints has he written into the image?

• •

"Action Flow, U.S. Carbon Dioxide Research and Assessment Program," in David H. Slade, Workshop Correspondences, 1979 AAAS/DOE Workshop Files, AAAS Archives.

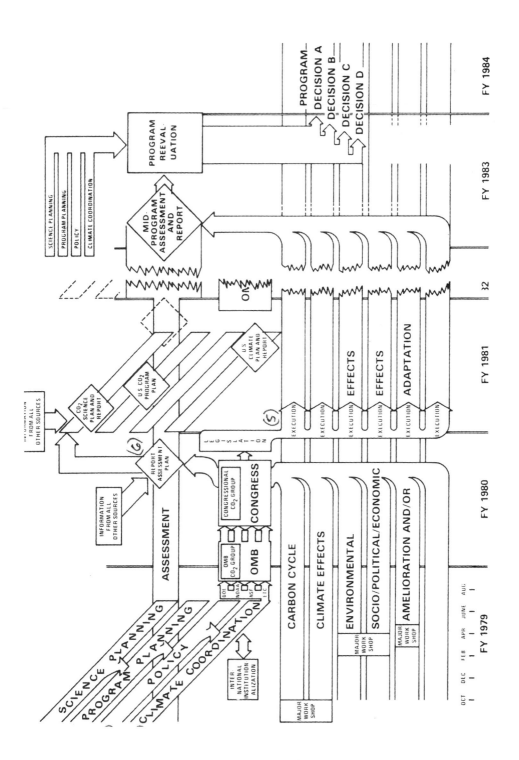

"ACTION FLOW"

DAVID SLADE

LETTER TO DAVID BURNS

The election of Ronald Reagan changed the landscape of climate science and climate change advocacy considerably. Whereas the Carter administration had signed the National Climate Program Act and collaborated with the American Association for the Advancement of Science (AAAS), Reagan's appointees sought to dismantle these kinds of programs, downplaying environmental risks across the board. Most famously, the new administration cut social science and environmental science research, including an 87 percent cut to the budget of the Department of Energy's new Solar Energy Research Institute in Boulder, Colorado. Much to the chagrin of some scientists and Democrats in Congress, the Reagan administration also targeted the AAAS-DoE CO_2 Project. A wry letter from the DoE's David Slade to the project liaison on the AAAS side, David Burns, captures the mood of the transition.

• •

David Slade to David Burns, October 2, 1980, AAAS Climate Program Records, AAAS Archives.

Department of Energy
Washington, D.C. 20545
October 2, 1980

Mr. David Burns, AAAS

Dear David:

I am not paranoid but, beside wrapping my head in silver foil before retiring so as to ward off the rays the CIA beams at me nightly, I carefully review publications sent to me to see how "they" are cutting me off from my information sources. Here is a typical example. As the high priest of CO_2 for the Federal government I view it as more than just a coincidence that my copy of *Science* is missing the article on CO_2.

   Hah; not so crazy after all, huh?

Slade

AL GORE

TESTIMONY BEFORE THE HOUSE COMMITTEE ON SCIENCE AND TECHNOLOGY, JULY 31, 1981

The administration's attack on social and environmental science research in the early 1980s did not go uncontested. In fact, a shared opposition to Reagan's policies on science and environmental protection brought climate scientists, environmentalists, and congressional Democrats together in a new way in the first half of the decade. Here, in a testy 1981 hearing, then representative from Tennessee Al Gore confronted Reagan's Department of Energy appointee, N. Douglas Pewitt, on the administration's efforts to slash the federal budget for carbon dioxide research, which had been approved under the National Climate Policy Act—the first in a series of increasingly hostile exchanges that helped associate climate change with other environmental issues championed by young congressional Democrats like Gore.

• •

Mr. PEWITT. Thank you, Mr. Chairman. It is a pleasure to be here.
    We submitted testimony obviously directed at the wrong subject. So I will just let that stand in the record and make a few comments reacting to what apparently seems to be the concern here. Several months ago, I testified in defense of a Department of

---

Al Gore, testimony in *Carbon Dioxide and Climate: The Greenhouse Effect*, Hearings before the House Committee on Science and Technology, 97th Cong. (1981). Excerpt: pp. 83–85, 87, 88.

Energy carbon dioxide program in which the administration proposed to spend $14 million in 1982, up from $9.7 million in 1981. This request was the same as the previous administration's.

Despite the hallway gossip we have heard here today, the budget request of the administration has not changed. We intend to use the funds in 1982 in the way that the congressional budget request asked to use those funds. I know of no policy change in the Department. There have been some managerial changes that we have undertaken in order to more effectively manage the Department's funds.

That concludes my statement. I have next to me the acting head of the carbon dioxide program, Dr. Koomanoff. We will try to respond to any technical questions or speak in general to the policy. Also present today is Dr. Kane who is head of the basic energy sciences program.

Mr. GORE. Dr. Pewitt, we have heard there has been a cut in the direction of the program and in the research included in this program. Is that inaccurate?

Mr. PEWITT. Absolutely. In order to cut it would be necessary to reprogram. We have proposed no reprograming or changed our budget request to the Congress.

Mr. GORE. Fine, go ahead. . . .

Mr. PEWITT. There are some changes being made in the management of the program. One of the changes is, we will not ask people for advice on how to put a program plan together and turn right around and feed them money that they helped formulate the program plan for. That is not the way to do the public business. We are not going to do that. We are not going to do social research on how Congress makes their decisions under the rubric of the carbon dioxide program. However, the program that we are going to execute is the program set forth in our congressional budget request in both January and in March.

Mr. GORE. Let me just ask you, in fiscal year 1982 in the area of effects of climate change and carbon dioxide increase on the environment, you still plan to spend $3.324 million?

Mr. PEWITT. That is not part of our congressional budget request. We intend to spend an appropriate amount of money on those environment effects. I think we have already let a contract this year for $700,000 with the Department of Agriculture to start looking at some of those effects.

Mr. GORE. Let me clarify this in my own mind. This is not hallway gossip, this is the National Climate Program Office document that says "Prepared by the United States Department of Energy Assistant Secretary for Environment." And it says, "In the program summary, effects of climate change and carbon dioxide increase on the environment, $3.324 million."

Mr. PEWITT. That is not part of the Department's budget request.

Mr. GORE. That is what the Department says is going to be spent on that area of carbon dioxide assessment effort and my question to you is, has that been reduced?

Mr. PEWITT. We are addressing that now. We are presently assessing this program and no commitment has been made to a specific level of funding. As yet, I do not know what the right level of funding is. Our congressional budget request does not specify those specific amounts.

Mr. GORE. The national climate program report did, the Department of Energy publication.

Mr. PEWITT. And the publication date on that was—

Mr. GORE. This is January of 1981.

Mr. PEWITT. Yes, sir, I am aware of that. However, that is not a departmental budget request. I have never seen the document and that is not what I testified in defense of.

Mr. GORE. But that was an accurate reflection of what the Department's program was planning to spend in this category?

Mr. PEWITT. Some months ago, that is correct.

Mr. GORE. And that is now being reduced, isn't it?

Mr. PEWITT. That is being reviewed; not reduced.

Mr. GORE. It may be increased?

Mr. PEWITT. I do not know. The director of the program, in whom I have absolute confidence, is looking into it. He and I have been in very close touch and while we are addressing the issue seriously, we do not know as yet what the right amount is to execute a program.

Mr. GORE. Let me go to the next item, social, political and economic costs and/or benefits of the global environment change.

That figure is $960,000. Do you still plan to spend that amount—

Mr. PEWITT. We plan to review that amount too and see if it is sensible.

Mr. GORE. Maybe the staff of Mr. Scheuer's subcommittee picked up information informally, what you refer to as hallway gossip. But that source of information indicated this part of the program was in some jeopardy. Is that inaccurate?

Mr. PEWITT. When I found out that, in the social research part of this, there was research aimed at things like how Congress makes decisions, I was less than happy that we would propose, in the rubric of carbon dioxide assessment, to spend the taxpayers' money on that sort of social research. That is appropriate at some point perhaps but not appropriate as part of the carbon dioxide program.

Mr. GORE. You may be right about that. What do you consider useful social and economic analyses in this area?

Mr. PEWITT. I have a program director currently looking at that, and he is going to come back to me with a proposal. We are planning to address the issues that, if addressed, can come up with something useful. We are not going to fund research under the rubric of carbon dioxide that possibly can't find a home someplace else....

Mr. GORE. I hope that the Department will not cut the research in this area and will include an appropriate amount for analysis of the economic and social effects of this problem, and will consider the possibility in reviewing the fiscal year 1983 request of doing everything possible to accelerate the effort to narrow these areas of uncertainty.

I take it that—let me ask you, do you share the sense of urgency that I feel about getting a firmer grip on whether or not this problem is in fact occurring?

Mr. PEWITT. No.

Mr. GORE. Tell me why?

Mr. PEWITT. I think that in running a scientific research program, we have a responsibility to the Congress and the American people to act in a fashion that is not alarmist. We have been down this road in similar areas before; for example, in aerosols and in the SST's [Supersonic Transport] disturbing of the atmosphere. It clearly is the prerogative of the Congress and policymakers to set a higher priority than the scientific programs can produce research results. There is a natural approach to scientific research that probably would not justify much of the more alarming carbon dioxide statements. I absolutely refuse as an official in a responsible position to engage in the type of alarmism for the American public that I have seen in these areas time and time again, and I do not think that I can responsibly encourage that sort of alarmism....

Mr. PEWITT. Understanding it is not easy. We have some bright people advising us on this program. We don't, as Federal officials, try to put our scientific judgment against others. We try to reach and get reasonable advice, and use peer review and advisory bodies to tell us how to proceed in this area. One should make sure that their advice is solidly based scientific advice.

Roger Revelle, the distinguished scientist that you had up here before, is part of that advisory body. You can't have bureaucrats dictating science, but at the same time you can't have scientists using alarmism in order to justify bigger research budgets. That is irresponsible, too. We are just trying to approach this issue reasonably.

RAFE POMERANCE

TESTIMONY BEFORE THE HOUSE COMMITTEE ON SCIENCE AND TECHNOLOGY, FEBRUARY 28, 1984

While congressional Democrats' relationships with Reagan appointees soured over research into climate change and other environmental issues, their relationships with American environmentalists became friendlier. In a 1984 iteration of the same series of hearings before the House Committee on Science and Technology, Gore hosted Friends of the Earth president Rafe Pomerance, whose views demonstrated at once an increasing interest by environmental groups in the climate change issue and a growing frustration with both scientists and politicians for their failure to act on evidence of CO_2-induced warming. The hearing took place shortly after the 1983 publication of two conflicting reports on climate change, one from the National Academy of Sciences and the other from the EPA, taken up in the next part.

....

Mr. POMERANCE. Thank you, Mr. Chairman. First, let me say a word of congratulations to you. I think that this topic is by far the most important environmental issue that there is. I think it has received too little consideration in the Congress, and I think that your leadership is probably the most important thing in the Congress that has come along on this issue. I have been working on

---

Carbon Dioxide and the Greenhouse Effect: Hearings before the House Committee on Science and Technology, 98th Cong. (1984). Excerpt: pp. 107–8, 108–9.

this issue for 7 years, since 1977, and I think that Congress, as an institution, sorely needs to pay much more attention to this problem.

Mr. GORE. Well, thank you very much. I want to make sure the reporter got all that. [*Laughter.*]

Thank you. Go ahead.

Mr. POMERANCE. I just have to avoid saying "Senator." [*Laughter.*]

Mr. GORE. We'll have equal time, equal time.

Mr. POMERANCE. Just a word about my own work on this. I worked on the Clean Air Act for many years in the midseventies, and after that did a good deal of reading on the CO_2 problem, and was named a member of the CO_2 Advisory Committee of the Department of Energy in the late seventies. The committee stopped meeting in the eighties. It was never disbanded, to my knowledge, but I always found the discussions were—I think I was the only person with a bachelor's degree in that forum—it was rather intimidating but very fascinating, because we debated and discussed all the most important issues, or at least the ones that people thought of at the time. I think it was unfortunate, but—

Mr. GORE. Are you saying the others were high school dropouts? [*Laughter.*]

Mr. POMERANCE. I figured that was coming.

Mr. GORE. Go ahead.

Mr. POMERANCE. Let me say a word, just in contrast to most of what has been said this morning. I think it is time to act. In fact, it's really too late to avoid, or it appears, initial warming. We know what to do. The evidence is in. The problem is as serious as exists. People talk about not leaving this to their grandchildren. I'm concerned about leaving this to my children. When we were working on the problem in the early seventies, you can see there has been a big bump just since the late seventies until now. The longer we wait, the more trouble we're going to be in, and this morning's testimony I will say did not leave me any easier with the notion that the oceans all of a sudden might change their takeup of CO_2. I believe this is a legacy we cannot leave to future generations. We have a fairly benign climate globally, and I don't think we should put it at risk.

The research budget on this issue is far too little. DOE has $13 million, and we know that $13 million is not very much money in terms of what Federal research dollars are pushed around. I feel that anything that the scientific community requests, just about, should be granted. I am not a scientist; I have no self-interest in that, but it is ridiculous to be spending such small sums of money on one of the most formidable problems that civilization faces.

A comment on the scientific community: Having listened to many of these hearings over the years, my conclusion is that we will never or virtually never hear from the scientific community in terms of telling you Members of Congress to act. You are the ones who are going to have to make that decision. Don't rely on the scientists. It's not their job. They're not going to tell you. They're going to say, "It's not my arena." . . .

Those are a few observations. I would just like to turn to our recommendations here and read those. This is the conclusion of our views, that some climatic change from fossil energy use, industrial pollution, and agricultural practices will occur. We urge prompt efforts to plan for adaption to these changes in low-lying areas. We support immediate funding for a large-scale global research program on ocean circulation. We understand that some consideration is being given to a joint EPA/NSF/NASA effort under the latter's proposed global habitability project, utilizing the space shuttle. As demonstrated by the recent El Niño effects, this research would be of great practical importance, far beyond the analysis of CO_2-caused climate warming.

We urge that energy efficiency and energy conservation become the top goal of U.S. domestic and foreign energy policies, in order to prevent the severe climatic changes that will occur if average world temperatures rise 4 degrees above current levels. Energy companies, and especially electric utilities, should be mandated to invest in energy conservation first and conventional energy production only second when applying for certificates of public need.

In this connection, we recommend abolishing the Synthetic Fuels Corporation and reappropriating funds to speed implementation of energy conservation and efficiency at the Federal, State, and local levels. Greatly increased research and development funds should be allocated to solar and renewable energy, including biomass.

Policies to ameliorate world climate change also facilitate solutions to other environmental problems, such as acid rain, while providing economically efficient energy investments and promoting energy security. Especially we recommend consolidation of research and development in this area to bring about more rapid policy change. We also urge immediate and large increases in research to study trace greenhouse gases other than CO_2.

U.S. support for conservation and renewable energy sources will stimulate greater acceptance of these resources by developing nations just when their energy use is expected to begin making a serious contribution to world CO_2 and trace greenhouse gas emissions. We urge that the recommitment of CO_2 climate change and other geochemical cycling efforts be incorporated as criteria for U.S. support of international development projects as well as our international science and technology policies.

We must act now to forestall serious climatic change beyond what we are already committed to. The United States bears a special responsibility as the world's largest user of fossil energy and the Nation with the world's second largest coal reserves. We should not continue man's experiment with world climate until we have a far better understanding of what the results will be. Today we do not know the consequences. Once we do, they will be irreversible for centuries. Furthermore, policies to limit climate change make the best dollars and sense.

Thank you.

PART 4

CLIMATE CHANGE AS CONTROVERSY

At the same time that environmentalists began to think about how to incorporate the problem of climate change more meaningfully into environmental advocacy, climate change discourse began to take on new political dimensions that made the issue more controversial, both within the scientific community and outside of it. In particular, a series of episodes in the 1970s and 1980s created rifts within the scientific community that would migrate into other, more public arenas, polarizing the entwined science and politics of global warming in ways that have affected climate change discourse ever since. The documents presented here revolve around three of these episodes. First, the debate between proponents of a global cooling hypothesis and a global warming hypothesis in the early 1970s revealed technical, disciplinary, and institutional rifts within the climate science community that would persist in the coming decades. Second, the publication of conflicting reports by the National Academy of Sciences and the Environmental Protection Agency (EPA) in 1979 and 1983 brought these scientific disagreements into public view.[1] And third, a concomitant debate over another potential anthropogenic climate event, nuclear winter, further helped to realign these increasingly bitter conflicts along partisan political lines. What was it about these experiences that helped foster the polarized, partisan, and often vitriolic discourse on climate change that has characterized climate change politics in the twenty-first century?

Each of three episodes in this chapter involved important and legitimate scientific disagreements, but in each case these disagreements also entwined with debates about the appropriate use and presentation of science in the public sphere. These debates strayed beyond the pages of peer-reviewed scientific publications into documents that straddled the line between scientific and public discourse. Scientists continued to discuss climatic change in journals like *Science*, but they also wrote and reviewed popular books, published scientific assessments at the behest of government agencies, and put out press releases, opinion pieces, and articles in nonscientific publications like *Foreign Affairs*.

Liminal scientific sources—that is, documents written by scientists about their subject matter outside of the usual peer-review structure of science, sometimes called gray literature—represent at once a trove and a minefield for historians of global warming. Once again, the difficulty of treating liminal sources often arises from the presentist paradox, the difficulty of taking historical documents on their own terms when our interest in them stems from present-day concerns.

Few episodes demonstrate the challenges of contextualizing nonscientific sources as well as the debate over the direction and severity of climate change in the 1970s. Historians and scientists revisiting the theory of global cooling characterize the episode in one of two ways. On one side, since the early 1990s, global warming skeptics have pointed to the unrealized warnings of a cooling climate and an approaching ice age that appeared in some scientific literature and frequently in the popular media during the 1970s, citing them as evidence of the unreliability of climate science. They use the historical "consensus" on global cooling to lampoon and undermine current consensus on global warming. On the other side, Thomas Peterson, William Connolley, and John Fleck have reviewed peer-reviewed literature from the period to reveal that the cooling thesis—a thesis espoused largely by more traditional climatologists working with historical evidence, who held that a combination of aerosols and dust from volcanic eruptions, human industry, and growing deserts could block radiation from the sun and thereby cool the earth—had relatively few adherents in the 1970s.[2] The "consensus" was thus a false one; history busts the myth.

Peterson and his colleagues are right to use scientific literature to undercut a simplified and invidious misreading of the past. At the same time, however, their conclusion masks the real and important divisions within the climate science community that liminal scientific literature on the global cooling debate can reveal. Both stories—the skeptical attack on the reliability of climate science and the affirmation of continued consensus—read the past as a way to support a modern political position, and in doing so, both groups distort the past. The story is in fact more complicated than a question of whether scientists believed in the potential of global cooling.

For example, take a look at the two CIA reports on climate change and food security from 1974 included in this chapter. According to the CIA, what institutional and disciplinary divisions characterized climate science during the 1970s? Why was the CIA interested in climate change during this period? For the CIA, which is more important, the severity and unpredictability of a changing climate or the direction of climate change? Comparing scientists' popular treatments of the subject—and other scientists' reviews of those treatments—reveals similar points of contention within the scientific community. In the liminal scientific literature—for example, Helmut Landsberg's review of Schneider's book, *The Genesis Strategy*—technical, disciplinary, and institutional debates over warming and cooling cut across a set of debates about what constitutes appropriate scientific behavior in the face of what some considered an imminent climatic threat. Apparently opposing scientists like Reid Bryson, a partisan of cooling, and Stephen Schneider, an early cooling theory adherent who saw flaws in his own models and later became vocal on global warming, had more in common with each other in their public concern over severe change than they did with more scientifically conservative peers in their own scientific camps.

Taken on their own terms, debates in the 1970s over the direction and severity of climate change involved disagreements over both scientific content and appropriate ways of conducting and communicating science. These same debates carried over into the 1980s, most notably and importantly in popular and scientific discussions of the potential climatic impacts of a nuclear war, commonly referred to as the "nuclear winter" episode.[3]

Coined by NASA's Richard Turco in 1983, the term *nuclear winter* describes a dramatic decrease in global mean temperatures that Turco and his collaborators—Brian Toon, Tom Ackerman, Jim Pollack, and most famously Carl Sagan (the TTAPS group for short)—believed would accompany a major nuclear exchange. Originally, the group sought to publish their findings in *Science*, but as a famous science popularizer and host of the hit television series *Cosmos*, Carl Sagan also sought to reach out to a broader audience to use the TTAPS study to support disarmament efforts. He published an exposé in the popular Sunday news supplement *Parade*; he helped host the Rockefeller Foundation–funded "Conference on the Long-Term Worldwide Biological Consequences of Nuclear War"; and he penned a detailed and specific proposal for a policy response to the nuclear winter findings in the journal *Foreign Affairs*, included here.[4]

What followed was a maelstrom of criticism, addendum, and qualified support that appeared in a diverse array of scientific and liminal scientific literature. A number of more conservative scientists—some "small c" scientific conservatives, others "big C" political Conservatives—attacked Sagan's policy recommendations, both by challenging the TTAPS group's scientific findings and by openly questioning the group's political motives, suggesting that their activism transcended the boundaries of good science. At the same time, however, a group of scientists led in part by Stephen Schneider at the National Center for Atmospheric Research (NCAR)—scientists who sometimes went out of their way to rebut conservative attacks on the TTAPS group—engaged in a series of debates of their own with Sagan's group over the specific content of the TTAPS report. In particular, after running new models, Schneider and his colleagues urged Sagan to back off of the "threshold concept" that underpinned his policy response. Like the original paper, the debate bled into the pages of more popular outlets like *Foreign Affairs*, where Schneider and Sagan were framed as adversaries. In the end, the debate over nuclear winter—both in the scientific and liminal scientific literature—helped foster the construction of a robust, standardized, and comparable interinstitutional computer model of global circulation, the Community Climate Model (now the Community Climate Systems Model), which would serve as a basic tool in climate science for the next three decades. And yet, the nuclear winter debate also helped polarize the climate science community politically in ways that have shaped the scientific and political discourse on climate change ever since.

As the cooling controversy of the 1970s and the nuclear winter debates of the 1980s demonstrate, liminal scientific literature is a broad but very important category. Consequently, reading it in context requires careful attention to form. Disagreements over scientific content and scientific propriety look different, for example, in book reviews in *Science* than they do in articles in *Foreign Affairs* or in conflicting scientific assessments by the National Academy of Sciences and the Environmental Protection Agency. These differences force us to ask questions about how the structure, purpose, and composition of a document—be it a book review, a popular essay, or a National Academy assessment—shape a document's content and the contemporary response to that content. Who is the intended audience of, say, a scientific press release or a popular article on a scientific issue? How do the writing conventions and standards of evidence change between, say, a book review for *Nature* and a National Academy assessment? How are the potential policy impacts of a study on nuclear winter published in *Science* different from the impacts of Sagan's *Foreign Affairs* article, and how might those differences affect an article's reception both among the broader public and within the scientific community? Thinking about the opportunities and limitations of different forms of liminal scientific literature enables us to explore the changing relationships between climate scientists and the broader public as the true stakes of the climate change issue began to unfold.

NOTES

1   Stephen Seidel, Dale L. Keyes, and United States Environmental Protection Agency Office of Policy and Resource Management Strategic Studies Staff, *Can We Delay a Greenhouse Warming? The Effectiveness and Feasibility of Options to Slow a Build-up of Carbon Dioxide in the Atmosphere*, 2nd ed. (1983); Carbon Dioxide Assessment Committee, *Changing Climate* (Washington, DC: National Academy of Sciences, 1983).
2   Sid Perkins, "Story One: Cooling Climate 'Consensus' of 1970s Never Was: Myth Often Cited by Global Warming Skeptics Debunked," *Science News* 174, no. 9 (October 2008): 5–6; Thomas C. Peterson, William M. Connolley, and John Fleck, "The Myth of the 1970s Global Cooling Scientific Consensus," *Bulletin of the American Meteorological Society* 89, no. 9 (September 2008): 1325–37.
3   For a complete account of the nuclear winter saga, see Lawrence Badash, *A Nuclear Winter's Tale: Science and Politics in the 1980s* (Cambridge, MA: MIT Press, 2009).
4   Paul R. Ehrlich et al., "Long-Term Biological Consequences of Nuclear War," *Science*, n.s., 222, no. 4630 (December 1983): 1293–1300.

U.S. CENTRAL INTELLIGENCE AGENCY

A STUDY OF CLIMATOLOGICAL RESEARCH AS IT PERTAINS TO INTELLIGENCE PROBLEMS

Between 1968 and 1972, a series of weather anomalies played havoc on the agricultural output of a number of key wheat- and rice-producing regions of the world, including India, much of Central Africa, and the Soviet Union. Extended droughts, failed monsoons, and untimely rains upset global markets and led to increases in domestic food prices and an embarrassing grain sale from the United States to its Cold War adversary, the Soviet Union.* Recognizing the potential geopolitical consequences of changes in weather and climate, the CIA launched two studies in 1974, the first to assess the intelligence issues involved in climate change and to articulate the state of climate science, the second to more specifically address food security issues in the 1970s. Though something of a caricature rather than a nuanced exploration of different climate methodologies—and one that failed to recognize the extent to which dynamic climate modelers like Joseph Smagorinsky had come to dominate the field by the 1970s—the CIA provided an interesting portrait of the landscape of climate science in the early 1970s. Despite the prevalence of CO_2-induced warming in the literature and in the models,

---

U.S. Central Intelligence Agency, Office of Research and Development, "A Study of Climatological Research as It Pertains to Intelligence Problems" (Washington, DC: U.S. Government Printing Office, 1974). Excerpts: pp. 1–3, 5, 6, 15, 16, 19, 21, 25–27.

* At the time the media referred to the exchange as the "Russian Wheat Deal," but it soon became known as "the Great Grain Robbery." See James Trager, *The Great Grain Robbery* (New York: Ballantine Books, 1975).

however, the CIA seems to have ignored warming altogether in favor of more popular fears about a return to the Little Ice Age. Why did the CIA care about climate change in 1974? And why, based on their priorities here, might their analysts favor a cooling hypothesis?

• •

The western world's leading climatologists have confirmed recent reports of a detrimental global climatic change. The stability of most nations is based upon a dependable source of food, but this stability will not be possible under the new climatic era. A forecast by the University of Wisconsin projects that the earth's climate is returning to that of the neo-boreal era (1600–1850)—an era of drought, famine, and political unrest in the western world.

A responsibility of the Intelligence Community is to assess a nation's capability and stability under varying internal or external pressures. The assessments normally include an analysis of the country's social, economic, political, and military sectors. The implied economic and political intelligence issues resulting from climatic change range far beyond the traditional concept of intelligence. The analysis of these issues is based upon two key questions:

Can the Agency depend on climatology as a science to accurately project the future?

What knowledge and understanding is available about world food production and can the consequences of a large climatic change be assessed?

Climate has not been a prime consideration of intelligence analysis because, until recently, it has not caused any significant perturbations to the status of major nations. This is so because during 50 of the last 60 years the Earth has, on the average, enjoyed the best agricultural climate since the eleventh century. An early twentieth century world food surplus hindered U.S. efforts to maintain and equalize farm production and incomes. Climate and its effect on world food production was considered to be only a minor factor not worth consideration in the complicated equation of country assessment. Food production, to meet the growing demands of a geometrically expanding world population, was always considered to be a question of matching technology and science to the problem.

The world is returning to the type of climate which has existed over the last 400 years. That is, the abnormal climate of agricultural-optimum is being replaced by a normal climate of the neo-boreal era.

The climate change began in 1960, but no one including the climatologists recognized it. Crop failures in the Soviet Union and India during the first part of the sixties were attributed to the natural fluctuation of the weather. India was supported by massive U.S. grain shipments that fed over 100 million people. To eat, the Soviets slaughtered their livestock, and Premier Nikita Khrushchev was quietly deposed.

Populations and the cost per hectare for technological investment grew exponentially. The world quietly ignored the warning provided by the 1964 crop failure and raced to keep ahead of a growing world population through massive investments in energy, technology, and biology. During the remainder of the 1960s, the climate change remained hidden in those back washes of the world where death through starvation and disease was already a common occurrence. The six West African countries south of the Sahara, known as the Sahel, including Mauretania, Senegal, Mali, Upper Volta, Niger, and Chad, became the first victims of the climate change. The failure of the African monsoon beginning in 1968 has driven these countries to the edge of economic and political ruin. They are now effectively wards of the United Nations and depend upon the United States for a majority of their food supply.

Later, in the 1970s one nation after another experienced the impact of the climatic change. The headlines from around the world told a story still not fully understood or one we don't want to face, such as:

- Burma (March 1973)—little rice for export due to drought
- North Korea (March 1973)—record high grain import reflected poor 1972 harvest
- Costa Rica and Honduras (1973)—worst drought in 50 years
- United States (April 1973)—"flood of the century along the Great Lakes"
- Japan (1973)—cold spell seriously damaged crops

- Pakistan (March 1973)—Islam planned import of U.S. grain to off-set crop failure due to drought
- Pakistan (August 1973)—worst flood in 20 years affected 2.8 million acres
- North Vietnam (September 1973)—important crop damaged by heavy rains
- Manila (March 1974)—millions in Asia face critical rice shortage
- Ecuador (April 1974)—shortage of rice reaching crisis proportion; political repercussions could threaten its stability
- USSR (June 1974)—poor weather threatens to reduce grain yields in the USSR
- China (June 1974)—droughts and floods
- India (June 1974)—monsoons late
- United States (July 1974)—heavy rain and droughts cause record loss to potential bumper crop.

During the last year every prominent country has launched a major new climatic forecasting program.

- USSR reorganized their climatic forecasting groups and replaced the head of the Hydrometeorological Service.
- Japan is planning to launch a major earth synchronous (U.S. manufactured) meteorological satellite and has secured complete collection, processing, and analysis systems from the U.S.
- China has made major purchases of meteorological collection and analysis equipment from western industrial sources.
- India is studying the application of climatic modification to secure a more homogeneous distribution of moisture from an erratic drought/flood monsoon.
- The U.S. National Academy of Sciences is preparing its recommendation for a National Climatic Research Program.
- The National Science Foundation (NSF) and the National Oceanographic and Atmospheric Agency (NOAA) have developed a National Climate Plan which will be presented to the Office of Management and Budget (OMB) for funding in FY 76.

Climate is now a critical factor. The politics of food will become the central issue of every government. On July 19, 1974, the Kiev Domestic Service reported massive rains and quoted an old proverb from Lvov Oblast, "The rains come not on the day for which we pray but only when we are making hay." The climate of the neo-boreal time period has arrived.

WHAT ARE THE INTELLIGENCE ISSUES?

In 1972 the Intelligence Community was faced with two issues concerning climatology:

- No methodologies available to alert policymakers of adverse climatic change
- No tools to assess the economic and political impact of such a change....

Since 1972, the grain crisis has intensified. Each year the world consumes approximately 1.2 billion metric tons. Since 1969 the storage of grain has decreased from 600 million metric tons to less than 100 million metric tons—a 30-day world supply.

With global climatic-induced agricultural failures of the early 1970s, the stability of many governments has been seriously threatened. Many governments have gone to great lengths to hide their agricultural predicaments from other countries as well as from their own people. It has become increasingly imperative to determine whether 1972 was an isolated event or—as the climatologists predicted—a major shift in the world's climate....

Timely forecasting of climate and its impact on any nation is vital to the planning and execution of U.S. policy on social, economic, and political issues. The new climatic era brings a promise of famine and starvation to many areas of the world. The resultant unrest caused by the mass movement of peoples across borders as well as the attendant intelligence questions cannot be met with existing analytical tools. In addition, the Agency will be faced with tracing and anticipating climate modification undertaken by a country to relieve its own situation at the detriment of the United States. The implication of such a modification must be carefully assessed. . . .

Current Approaches to Climatology

There are three basic schools or philosophies of climatology. The first is centered around Professor H. H. Lamb, who is currently the Director of the Climatic Research Unit at the University of East Anglia in the United Kingdom. This school contends that if a climatologist is to project future climates, he must understand what has occurred in the past. The second is characterized by Dr. Joseph Smagorinsky, who is the Director of the Geophysical Fluid Dynamics Laboratory at Princeton University. This center believes that a complete understanding of atmospheric circulation is sufficient for climatic forecasting. The third is best represented by Dr. M. I. Budyko, an eminent Soviet climatological theoretician. He pursues the hypothesis that an understanding of the total distribution of thermal energy is necessary for climatic forecasting.

The Lambian school is based on the establishment of climatic statistical trends. A great deal of effort has been expended by the followers of this philosophy in quantifying the qualitative descriptions provided by historical sources (ancient court scribes, ship's logs, and scholars). Their reconstruction of climatic conditions has reached back 5,000 years. This particular approach was almost totally dependent upon manmade records. In recent years, the use of geophysical indicators such as tree rings, sedimentary deposits, and Arctic ice layering has added substantially to the global data pool. . . .

The Smagorinsky-ian school of climatology is based upon the meteorologist's attempts to extend the predictive capabilities of the equations of fluid motion. Meteorology deals principally with the forecasting of atmospheric pressure differentials and the propensity for given patterns to result in rain, snow, ice, high winds, etc. It does not take into account solar or Earth radiation nor hydrological (i.e., evaporation) variables.

Since the availability of serial, numerical computers in the latter half of the 1940s, the meteorologist has developed a system of models to predict near-term atmospheric variations. The basic tool employed by this group is the General Circulation Model. These models describe the effects of large-scale atmospheric motion and are treated explicitly by numerical integration. For almost 30 years the meteorologist has tried unsuccessfully to extend his predictive capability past a 24-hour forecast. The Smagorinsky-ian approach, however, is the currently accepted methodology within the United States Government and receives more than 90 percent of all the research and development funding available therein.

The *Budyko-ian* school is based upon the theoretical work of Dr. M. I. Budyko, who is associated with the Global Meteorological Institute in Leningrad. The basis of this approach to the climatological problem is Dr. Budyko's 1955 paper entitled, "The Heat Balance of the Earth's Surface." This paper advances the hypothesis that all atmospheric motions are dependent upon the thermodynamic effect of a nonhomogeneous distribution of energy on the Earth's surface. Though this work originally met with opposition from the world's meteorologists, it is now accepted as a more reasonable basis for developing a successful climatic prediction model. The earlier, simplistic explanation of climate was basically Budyko-ian.

RECENT MILESTONES

Explanation of the scientific phenomena and elaboration of the three methodological schools provide a background for more recent developments—developments which have more relevancy to requirements as they might emerge in the Intelligence Community. The University of Wisconsin's work appears to be providing the cohesion for continuing research in this area.

The Wisconsin Study

The University of Wisconsin was the first accredited academic center to forecast that a major global climatic change was underway.... They observed that the climate we have enjoyed in recent decades was extremely favorable for agriculture. During this period, from 1930 to 1960, the world population doubled, national boundaries were redrawn, the industrial revolution became a worldwide phenomenon, marginal lands began to be used in an effort to feed a vastly increased population, and special crop strains optimally suited to prevailing weather conditions were developed and became part of what was called the "green revolution."

The climate of the 1800s was far less favorable for agriculture in most areas of the world. In the United States during that century, the midwest grain-producing areas were cooler and wetter, and snow lines of the Russian steppes lasted for longer periods of time. More extended periods of drought were noted in the areas of the Soviet Union now known as the new lands. Moreover, extensive monsoon failures were common around the world, affecting in particular China, the Philippines, and the Indian subcontinent.

The Wisconsin analysis questioned whether a return to these climatic conditions could support a population that has grown from 1.1 billion in 1850 to 3.75 billion in 1970. The Wisconsin group predicted that the climate could not support the world's population since technology offers no immediate solution. . . .

It has been shown that over the last 10,000 years there have been many climatic changes of regional and global significance. Detailed descriptions exist showing how these climatic changes affected the people of these regions. The Wisconsin forecast suggests that the world is returning to the climatic regime that existed from the 1600s to the 1850s, normally called the neo-boreal or "Little Ice Age." . . .

For 250 years most of the world suffered major economic and political unrest which could be directly or indirectly attributed to the climate of the neo-boreal era. The great potato famine of 1845 in Ireland was the last gasp of the "little ice age." Yet for every death in Ireland there were ten in the Asian countries.

What would a return to this climate mean today? Based on the Wisconsin study, it would mean that India will have a major drought every four years and could only support three-fourths of her present population. The world reserve would have to supply 30 to 50 million metric tons of grain each year to prevent the deaths of 150 million Indians. China, with a major famine every five years, would require a supply of 50 million metric tons of grain. The Soviet Union would lose Kazakhstan for grain production thereby showing a yearly loss of 48 million metric tons of grain. Canada, a major exporter, would lose over 50 percent in production capability and 75 percent in exporting. Northern Europe would lose 25 to 30 percent of its present product capability while the Common Market countries would zero their exports.

A limited number of people within the United States are involved in climatological research. On the West Coast there are two significant groups. The first is under Dr. Larry Gates at the RAND Corporation in Santa Monica. Dr. Gates' work has been supported by ARPA and is theoretically Smagorinsky-ian. He has worked for three years under an ARPA grant utilizing basically the UCLA two-level General Circulation Model. Though the work has been theoretically interesting and has developed many new software capabilities, they have still not arrived at an operational system. Dr. Gates has been strongly impressed by developments in the Budyko-ian school and is in the process of modifying their simulation programs to incorporate some of the more recent thermodynamic developments.

The Scripps Institution group at La Jolla, under the direction of Dr. John Isaacs, and more recently with the inclusion of Dr. Jerome Namais, has followed both the Lambian and Budyko-ian approaches to climatological problems. Their main capabilities have been in the development of climatological observables. Dr. Isaacs' early work, which has been continued by Namais' research, was directed at the thermodynamic influence of the oceans on world atmospheric circulation. At present, no pragmatic climatological forecasting is being pursued at Scripps.

The atmospheric sciences group at the University of Arizona is solidly Budyko-ian. Dr. William Sellers who heads this group is one of the country's leading technicians in Budyko-ian methodology. His first published climatic model in 1968 was not well received by the Smagorinsky-ians or by the Budyko-ians within the world community. They did acknowledge, however, that it was the first pragmatic systematizing of this approach. His latest model, developed in 1972, has had a significant effect in crystalizing this whole philosophy and demonstrating a pragmatic climatological model.

There are two climate groups in the midwest—one being NCAR (National Center for Atmospheric Research) at Boulder, Colorado. Their efforts have been to explore highly disaggregated atmospheric models. The second group, at the University of Wisconsin, is under Reid Bryson and John Kutzbach, both mentioned earlier. Their work at Wisconsin represents the focal point for climatological research in the United States. They are the only people within the academic community in the United States that have a seasonal climatological forecasting system.

The eastern establishment, consisting of Princeton and the Massachusetts Institute of Technology, is primarily Smagorinsky-ian. They are basically NOAA-funded and, though primarily engaged in increasing the accuracy of meteorological forecasts, have attempted without success to provide climatological forecasting capabilities.

In summation, the eastern schools have employed basically the Smagorinsky-ian principles in one way or another. The limitation of this approach, although not yet apparent to the establishment, is rapidly being abandoned by the academic community. The pragmatic capabilities of the Budyko-ians and the methodologies therein are quickly being absorbed by both the East and West Coast establishments. The Lambians and their primarily statistical approach are beginning to lose favor, but their development of historical climatological records has provided a vital service within the climatological community.

S. I. RASOOL AND S. H. SCHNEIDER

ATMOSPHERIC CARBON DIOXIDE AND AEROSOLS

Effects of Large Increases on Global Climate

In the late 1960s and early 1970s, the climate science community focused much of their efforts on modeling two potential scenarios of climatic change: CO_2-induced warming and aerosol cooling. Whereas increased atmospheric CO_2 could absorb and re-release more heat within the climate system, increases in reflective aerosols like dust and volcanic ash might act to cool the earth by reflecting energy from the sun back into space. In 1970, most atmospheric scientists understood that both mechanisms were in play. In 1971, however, Ishtiaque Rasool and Stephen Schneider began to see that industrial pollution could add to the reflective aerosols contributed by volcanoes and agriculture, tipping the balance toward a runaway cooling that, in their words, might trigger another ice age. Published in *Science*, their paper on CO_2 and aerosols caused a major stir in the climate science community. Schneider later found an important error in the paper, which used a one-dimensional model of the atmosphere that accounted poorly for atmospheric circulation, and he quickly revised his conclusions. Even so, the paper—and the debate over warming and cooling—would haunt the credibility of

---

S. I. Rasool and S. H. Schneider, "Atmospheric Carbon Dioxide and Aerosols: Effects of Large Increases on Global Climate," *Science*, n.s., 173, no. 3992 (July 1971): 138–41. Reprinted with permission from AAAS. Schneider recalls the research and the responses to the paper in entertaining and revealing detail in *Science as a Contact Sport: Inside the Battle to Save Earth's Climate* (Washington, DC: National Geographic, 2009), 14–21.

climate scientists concerned about global warming for decades. How might the paper itself raise the hackles of scientists? How does Schneider's contribution to science journalism—a form of liminal scientific source—help frame his scientific paper with Rasool?

· ·

Abstract. Effects on the global temperature of large increases in carbon dioxide and aerosol densities in the atmosphere of Earth have been computed. It is found that, although the addition of carbon dioxide in the atmosphere does increase the surface temperature, the rate of temperature increase diminishes with increasing carbon dioxide in the atmosphere. For aerosols, however, the net effect of increase in density is to reduce the surface temperature of Earth. Because of the exponential dependence of the backscattering, the rate of temperature decrease is augmented with increasing aerosol content. An increase by only a factor of 4 in global aerosol background concentration may be sufficient to reduce the surface temperature by as much as 3.5° K. If sustained over a period of several years, such a temperature decrease over the whole globe is believed to be sufficient to trigger an ice age.

The rate at which human activities may be inadvertently modifying the climate of Earth has become a problem of serious concern. In the last few decades the concentration of CO_2 in the atmosphere appears to have increased by 7 percent. During the same period, the aerosol content of the lower atmosphere may have been augmented by as much as 100 percent. How have these changes in the composition of the atmosphere affected the climate of the globe? More importantly, is it possible that a continued increase in the CO_2 and dust content of the atmosphere at the present rate will produce such large-scale effects on the global temperature that the process may *run away*, with the planet Earth eventually becoming as hot as Venus (700°K) or as cold as Mars (230°K)?

We will report here on the first results of a calculation in which separate estimates were made of the effects on global temperature of large increases in the amount of CO_2 and dust in the atmosphere. It is found that even an increase by a factor of 8 in the amount of CO_2, which is highly unlikely in the next several thousand years, will produce an increase in the surface temperature of at least 2°K. However, the effect on surface temperature of an increase in the aerosol content of the atmosphere is found to be quite significant. An increase by a factor of 5 in the equilibrium dust concentration in the global atmosphere, which cannot be ruled out as a possibility within the next century, could decrease the mean surface temperature by as much as 3.5°K. If sustained over a period of several years, such a temperature decrease could be sufficient to trigger an ice age!

REID BRYSON

A PERSPECTIVE ON CLIMATE CHANGE

Climate scientists worked amid great uncertainty in the 1970s, but by 1974 many climatologists and modelers began to worry that their assumptions about the stability of global and regional climatic systems and the speed with which those systems changed were wrong. Here, University of Wisconsin climatologist Reid Bryson—a proponent of a theory of global cooling driven by aerosol reflection—articulates the key questions emerging in climatology in the middle of the decade in the journal *Science*. The article appeared in a peer-reviewed journal, but it was not a work of original scientific research. How does a literature review like this differ from, say, Rasool and Schneider's 1971 cooling paper? What purpose does a piece like this serve?

• •

In recent years, with the heightened concern about the impact of human activities on the environment and the more immediate concern about the impact of the environment on world food supplies, there has been an upsurge of interest in climatic change at all scales, from local to global. Discussions of climatic change in the general press and professional journals have ranged from completely theoretical to completely descriptive, and from detailed analysis of certain aspects of the problem to general speculation. However, there has been a dearth of discussion of climatic change from an historical perspective and minimal attention to certain important questions, such as: (i) How large

---

Reid Bryson, "A Perspective on Climate Change," *Science* 184, no. 4138 (May 1974): 753–60. Excerpt: pp. 753, 759. Reprinted with permission from AAAS.

must a climatic change be to be important? (ii) How fast can the climate change? (iii) What are the causal parameters, and why do they change? (iv) How sensitive is the climate to small changes in the causal parameters? This article is an attempt to provide an element of perspective on these questions. . . .

At the beginning of this article I posed certain questions about the size of important climatic changes and the magnitude of changes in the causal parameters needed to produce these changes. I showed that numerically small changes in climatic variables may produce significant environmental changes and that rather small changes in the extrinsic control variables are adequate to explain these responses. A significant feature of recent paleoclimatic research is that significant climatic pattern changes are surprisingly small—explanation of past environments does not require drastic modification of the general circulation.

I also suggested that several of the extrinsic control variables may be significantly modified by human activity. These include the turbidity of the atmosphere and its carbon dioxide content.

The data presented in several of the figures show that climate can change very rapidly. While some causal parameters, such as earth-sun geometry, change slowly, others, such as volcano-induced turbidity, may change rapidly and sporadically. Apparently the only controls of the speed of climatic change are the time constant of the active layer at the surface of the earth and the time constant of glaciers. Ice age climates may end (and probably start) in a century or two, although glacial and oceanic response and new equilibrium may take millennia. Holocene climatic changes, smaller in magnitude, may be accomplished in decades. The overriding present question, of course, is how the present climatic change will develop. In perspective, such changes do not appear to be random fluctuations from some long-term "normal."

STEPHEN H. SCHNEIDER

THE GENESIS STRATEGY

One of the more controversial pieces of liminal scientific literature on climate change in the late 1970s was Stephen Schneider's cautionary exposé *The Genesis Strategy*. The same scientist who had, with Ishtiaque Rasool, cautioned in 1971 that aerosol cooling might tip the climate system toward another ice age had by 1976 come around to the potential dangers of climatic change induced by CO_2 warming. For Schneider, it wasn't so much the direction of climate change that mattered—and in fact he was willing to entertain arguments for a variety of mechanisms of change—but the way that increased climatic variability created vulnerability in human and natural systems. Here, in the book's preface, Schneider explains his philosophy in terms of the biblical story of Joseph and explores what he sees as a problem of conservatism in the scientific community. Why start with the story of Joseph? What does the biblical reference say about his intended audience, and why might some scientists object to starting the book this way? Below, a series of exchanges between Schneider and one of the book's harshest critics, Helmet E. Landsberg, in the pages of the journal *EOS: Transactions of the American Geophysical Union*.

• •

Perhaps we can learn from a story in the Old Testament that seems especially applicable to such dilemmas. In the Book of Genesis, Joseph interpreted one of Pharaoh's dreams as a warning that seven years of famine were to follow the seven years of feast that Egypt would enjoy.

Stephen H. Schneider, *The Genesis Strategy: Climate and Global Survival* (New York: Plenum Press, 1976). Excerpt: pp. xii–xiv.

Pharaoh, choosing not to take the chance, heeded Joseph's advice and had food stored up against the possibility of a food shortage. Events happened as predicted. In this story we see the vindication of Joseph's warning against the famine "wolf" (which no doubt, is the best long-range weather forecast ever made) and of Pharaoh's action to protect his "flock" by storing food to hedge against the *possibility* of famine.

As the United States enters its third century as a nation, there are those who warn that the hungry wolf of biblical times is lurking in the woods again. They advise that we adopt the proven "Genesis Strategy" (as I dubbed Joseph's plan in an article in 1974) as the prudent response to the prospect of climate-induced famine.

However, the world has grown much more complex since Joseph's time. Agricultural technologies exist that were not available to the people of Genesis—tools such as modern irrigation systems, artificial fertilizers, genetic-hybrid seeds, insecticides, and more efficient and sophisticated methods of transporting and storing food, to name a few. Today, building and maintaining food reserves are expensive procedures that dramatically stabilize market prices once the reserves have been established; accordingly, many producers and most food traders resist adoption of this strategy, calling it an obstacle to profit incentives or at least unnecessary in today's technological world.

The Genesis Strategy of maintaining large margins of safety to secure our means of survival can be applied to many of the urgent questions of human survival. One of these questions is climatic change, a phenomenon that deserves wide attention for several reasons: (1) It is often global in scope and concretely demonstrates potentially high risks to world stability; such risks can result from shortsighted political practices that stress short-term national solutions to problems that are often both long-term and worldwide. (2) Issues of climatic change have been underplayed or incompletely treated in most debates about the world's survival. (3) Nuclear technology was unleashed on the world before enough people had adequate appreciation of its dangers to demand strict control, but there is probably still time left to anticipate and even to prevent a number of climate-related crises, in addition to a climate-food disaster, that may lie ahead.

To be sure, there is plenty of controversy and conflicting evidence surrounding each potential crisis. It is important to present opposing views, even though this may necessitate some technical details. But I am convinced that ill-founded certainty is far worse than a realistic appreciation of the confusing issues of human survival. The public must be better informed if it is to be politically wise about these issues.

Any reduction in the number and degree of uncertainties surrounding the scientific testimonies of honest experts will go a long way toward helping modern technological societies make the wisest decisions. We must recognize all the while, however, that not every sighting of an iceberg will be followed by a collision. In addition, we must realize that the research needed to reduce the uncertainties we confront will be difficult, time-consuming, and expensive.

But my chief concern for the future is political rather than scientific; it is that some wolves will attack long before we are certain enough of their existence to feel compelled to effect difficult political actions. Such actions, however expensive or unpleasant, may be vital to hedge against dangerous plausibilities long before scientific certainty about their magnitude and timing is established. Since most predictions of threats to our survival are by no means guaranteed to come true, I feel compelled to go into considerable detail on the scientific bases of a number of dangers, in order to show that concern may nevertheless be justified. Although I will attempt to be as objective as possible in discussing scientific evidence, I have been unable to avoid injecting my own personal philosophies into some of the discussions, particularly those addressing the question of whether present scientific evidence justifies immediate action. Realizing that total objectivity is impossible, I have tried throughout this book to state my biases openly and to help the reader separate personal or political philosophy from scientific opinions.

I am deeply convinced of the real dangers that societies face in the years ahead and the need to clarify the importance of the scientific component of these dangers. To convey my sense of urgency, I have decided to eschew the traditional role of the scientist to advance knowledge quietly, and instead to write a book that mixes politics and science and often goes beyond the confines of my academic training. I agree with John Kenneth Galbraith's advice to scientists (and to those others who claim the title of "intellectual") to step past the traditional, and comfortable, boundaries of familiar scholarship and deal publicly in the real world of politics—especially when politics impinge on one's own scientific expertise.

The world predicament, crucial as it is to our future survivability, often appears to be a problem that we are powerless to influence as individuals. However, much of the dilemma can be traced to our rationalizations that we are not technically qualified or significant enough to have any impact on the solution to our growing problems. Thus, we abdicate our personal involvement and influence to leaders and specialists, who are supposed to understand these bewildering issues "better" than we do. But *there is a personal message here* directed to everyone, and I have tried to weave it into these pages: Most of the crucial issues of human survival that will confront humanity over the next few decades will call for ethical and political value judgments—decisions on *how to act in the face of uncertainties*. In few cases will these decisions be based on issues clear enough to be decided easily by an input of scientific truths comprehensible to only a handful of specialists. Human value judgments are too important to be left exclusively to the experts. Don't underestimate the worth of your contribution.

· ·

HELMUT E. LANDSBERG

REVIEW: THE GENESIS STRATEGY— CLIMATE AND GLOBAL SURVIVAL

· ·

This book is introduced as a cooperative effort of a scientist and a writer. But it might be better characterized as the opinions of a scientist interpreted by a good writer in an easy flowing style. Let it also be said in advance that it is poorly illustrated with generally inferior graphics, except for two cartoons—banished to the end. Their footnotes and references are merrily mixed in the abominable modern fashion which

---

Helmut E. Landsberg, "Review: *The Genesis Strategy—Climate and Global Survival*," *EOS: Transactions of the American Geophysical Union* 57, no. 9 (September 1976): 634–35.

requires leafing back and forth. The format does not consider the reader's convenience....

The arguments are presented in three parts. The first is an overview of the intricate relations between climate, technology, and human survival. It deals also with the alleged effects of inadvertent global climate modification. This is followed by a discussion of the world food problem. Part two deals with the climate-related crises. These are considered in the framework of climatic history, theories of climate, and attempts to modify weather and climate. Interspersed is a chapter on U.S. food production and the policies of the U.S. Department of Agriculture. The last part covers the topic of climate change as a cause of the world predicament, implications for the world food production, and revised institutional structures to alleviate the perceived difficulties.

The simplistic basic thesis of all this (without the benefit of systems analysis) is that because of climatic changes, or at least fluctuations, food reserves should be built up to keep the world's teeming masses from starving. The principle was first proposed by Joseph to the Pharaoh (Genesis 41:36). This famous story gave the book its title. There are, of course, many others, including the reviewer, who have advocated the same thing....

There are many valid points in Schneider's discussion. Among them is the inescapable conclusion that air and thermal pollution has to be controlled because of the potential for unwanted regional, if not global, climatic consequences. Similarly, warnings about overtaxing the ecosystem, locally or globally, cannot be repeated often enough. Nor should weather or climate modification be undertaken unless the outcome can be predicted and all ramifications assessed reliably in advance. Also, the proposals to convert tropical rain forests into arable land are rightly opposed in this book as a potential ecological disaster....

The wide-ranging potpourri of science, nature, and politics is multidisciplinary, as promised, but it is also very undisciplined. The author never comes to grips with really basic issues....

Much of the background for the book has been gathered from the press. Citations of articles from daily papers and magazines abound. This seems to me to be a very haphazard way to achieve a balanced view of an extremely complex problem. Similarly, in the presentation of climatological and meteorological problems many unbalanced statements appear. Conjectures are elevated to facts. Inadequate information is clothed in the mantle of scientific authority. In some instances, outrightly flimsy ideas are advanced paradigmatically to make a point. The meteorological causes of the recent Sahel drought are such a case.

One misses in the technical part of the treatise the dispassionate and critical attitude which has been such a distinguishing attribute of scientists. In particular, I disagree with the opinion that uncertain predictions derived from clearly inadequate mathematical-numerical models should be used for public (political) decisions. Nothing could erode the credibility of scientists faster than that. This does not mean one has to have absolute certainty, but one needs to give the decision makers a precise estimate of the uncertainty. Certainly back-of-envelope calculations may lead to dangerously false conclusions.

Finally, Schneider wants to reorganize the government and the world organizations. He makes a plea for a fourth branch in the national government, a superplanning board which in science and technology questions could apparently independently reverse policies of the executive branch. It is a bit vague and unbelievably naïve. . . .

On the world scene he pleads for four world security institutes: one dealing with "Imminent Disasters," another with "Resource Availability," the third with "Alternate Technologies," and the last exploring "Policy Options." The reader is not informed about the many existing national and international organizations which already work along these lines. One might just mention the International Red Cross, the World Health Organization, the Food and Agricultural Organization, the Economic and Social Council, and the U.N. Environmental Program. One may criticize these groups as being slow and cumbersome. Yet they have done a great deal of good work and are generally steering in the right direction. . . .

Schneider advocates that scientists run for public office. Perhaps he will try it. If he should, however, intend to contribute to further advancement of science, one might suggest that he spend less time going to the large number of meetings and workshops that he seems to frequent and also that he change his reading habits from newsprint—all through the simple expedient of reading scientific journals and otherwise being a regular devotee of a first class scientific library.

> Helmut E. Landsberg, director of the Institute for Fluid Dynamics and Applied Mathematics at the University of Maryland, is a past president of AGU.

•••

STEPHEN H. SCHNEIDER AND HELMUT E. LANDSBERG

FORUM

•••

COMMENT

Helmut E. Landsberg used his review... of my book, *The Genesis Strategy: Climate and Global Survival* (with Lynn E. Mesirow, Plenum, 1976), as a platform for a harangue on my professional values and personal style. Instead of keeping within the bounds of a substantive critique of the issues of climate and global survival raised in the book, he indulges in conjectures on such topics as how he thinks I spend my time in professional pursuits and speculates on how I might reform my ways if I "intend to contribute to further advancement of science." Ironically, to use Landsberg's own words, these "conjectures are elevated to facts." In any case, they have nothing to do with reviewing a book, and they prompted this rebuttal to set the record straight.

Moreover, when reverting to the appropriate role of book reviewer, he misstated several important aspects of the book's content. In particular, take his comment, "Schneider advocates that scientists run for

---

"Forum," *EOS: Transactions of the American Geophysical Union* 58, no. 3 (March 1977): 122.

public office. Perhaps he will try it." My very different views are stated plainly in the book (page 331). Furthermore, Landsberg describes my proposal for a fourth branch of government as an idea to create "a superplanning board in which science and technology questions could apparently independently *reverse* [my emphasis] policies of the executive branch." Yet (as evidenced on page 47 and again on page 307) no such role is suggested. Rather, I say that "the fourth branch would not have any legislative, administrative, or judicial powers; these rightly belong to the original three branches. In fact, to preserve the odd number of branches with decision making power, the fourth branch would have no power at all, *except* to gather and disseminate information."

Landsberg finally raises, but quickly dismisses, a number of specific science/public interface issues: for example, (1) the potential role in public policy making of state-of-the-art mathematical models for problems of human impact on climate (issues where there exists no precedent in climatic history and empirical studies are thus of insufficient guidance); (2) the use of newspapers or magazines as source material to quote public figures (who generally do not publish in scientific journals) or to describe very recent events; and (3) the role of scientists in a public arena. Although I disagree vigorously with many of his views, I appreciate Landsberg's touching on these points, since I wrote the book in large measure to generate widespread debate within the scientific community on just such issues. I am gratified that EOS has helped to foster such a debate and sincerely hope that the dialog can be continued, but at a level free from the polemics of personal attack.

Stephen H. Schneider
National Center for Atmospheric Research

REPLY

Because Schneider seems to have the potential of becoming a first-class scientist, a bit of advice appeared not out of place. He has the right to disagree.

His objections to my interpretations of what he wrote seem to be a result of the ambiguity of the statements. Take the proposed fourth branch of government. We read on page 47, "Although the objectives of their [Pirages and Ehrlich] Planning Branch and the Truth and Consequences Branch proposed here *are indeed similar*, I envisage ... that the latter might *also* operate primarily as a source of public information ... " and again on page 310 he states, "The fourth branch proposed here, although similar in many important aspects to the Planning Branch [proposed by Tugwell], differs *perhaps* in emphasis it places on public education" [my emphasis]. To me this reads like a planning organization with a lot of public relations fanfare.

His statement on running for public office is equally ambiguous. He only believes that scientists in legislative positions would quickly become obsolescent.

On the most important of his objections to my review, namely the use of climate models for the predictions to guide public policy, let me quote a more competent judge than I am. Joseph Smagorinsky in his excellent review "Global Atmospheric Modeling and the Numerical Simulation of Climate" (*Weather and Climate Modification*, John Wiley, 1974) states in relation to various policy decisions advocated to counteract potential man-made effects on climate, "Crude or premature estimates can be very misleading in providing guidance for such far-reaching decisions and may be far more damaging than no estimate at all" (page 673), and also, "It does warn us, however, that if current physically comprehensive models are inadequate to answer some of our questions, then certainly we should be wary of basic broad national or international decisions on hand-waiving arguments or back-of-the-envelope calculations" (page 685).

Not only do I fully subscribe to this assessment but it was the nub of my critique of the core of Schneider's book.

H. E. Landsberg
University of Maryland

NATIONAL ACADEMY OF SCIENCES

CARBON DIOXIDE AND CLIMATE

A Scientific Assessment

Climate science in the 1970s was characterized by a combination of great concern and great uncertainty. In 1979, as part of an effort to reduce the uncertainty and get a better handle on the concern, President Carter's Office of Science and Technology Policy requested that the National Academy of Sciences assess the state of the science on the CO_2 question. A private nonprofit scientific organization created by an act of Congress in 1863, the National Academy of Sciences (NAS) is meant to provide independent, objective advice on scientific issues pertaining to matters of science and technology. Characteristic of National Academy studies, the Climate Research Board's assessment in 1979 was not an attempt to create new scientific knowledge, but rather an effort to catalog what was known and what was still unknown on the CO_2 question. Often called the Charney Report after the committee's chair, Jule Charney, the assessment's deep and abiding concern over the impact of CO_2-induced climate change—and its establishment of a "best guess" of 3°C warming ±1.5° for a doubling of atmospheric CO_2—set the terms of the debate for the 1980s. How does the tone of the NAS assessment here differ from that of Schneider's book above? How would you compare it to the style of a scientific paper? Where is the NAS

Climate Research Board, "Carbon Dioxide and Climate: A Scientific Assessment" (Washington, DC: National Academy of Sciences, 1979). Excerpt: pp. vii–x, 1–2, 3. Reprinted with permission from the National Academy of Sciences, Courtesy of the National Academies Press, Washington, D.C.

mandate reflected in the text, and how does that impact the presentation of climate change here?*

· ·

FOREWORD

Each of our sun's planets has its own climate, determined in large measure by the planet's separation from its mother star and the nature of its atmospheric blanket. Life on our own earth is possible only because of its equable climate, and the distribution of climatic regimes over the globe has profoundly shaped the evolution of man and his society.

For more than a century, we have been aware that changes in the composition of the atmosphere could affect its ability to trap the sun's energy for our benefit. We now have incontrovertible evidence that the atmosphere is indeed changing and that we ourselves contribute to that change. Atmospheric concentrations of carbon dioxide are steadily increasing, and these changes are linked with man's use of fossil fuels and exploitation of the land. Since carbon dioxide plays a significant role in the heat budget of the atmosphere, it is reasonable to suppose that continued increases would affect climate.

These concerns have prompted a number of investigations of the implications of increasing carbon dioxide. Their consensus has been that increasing carbon dioxide will lead to a warmer earth with a different distribution of climatic regimes. In view of the implications of this issue for national and international policy planning, the Office of Science and Technology Policy requested the National Academy of Sciences to undertake an independent critical assessment of the scientific basis of these studies and the degree of certainty that could be attached to their results. . . .

* For a brief account of Charney's scientific life, see Norman Phillips, "Jule Gregory Charney, Jan. 1, 1917–July 16, 1981," *National Academy of Sciences Biographical Memoirs*, n.d., www.nasonline.org. For a detailed study of how the NAS and other bodies have assessed climate models and their results, see Jeroen van der Sluijs et al., "Anchoring Devices in Science for Policy: The Case of Consensus around Climate Sensitivity," *Social Studies of Science* 28, no. 2 (April 1998): 291–323.

The conclusions of this brief but intense investigation may be comforting to scientists but disturbing to policymakers. If carbon dioxide continues to increase, the study group finds no reason to doubt that climate changes will result and no reason to believe that these changes will be negligible. The conclusions of prior studies have been generally reaffirmed. However, the study group points out that the ocean, the great and ponderous flywheel of the global climate system, may be expected to slow the course of observable climatic change. A wait-and-see policy may mean waiting until it is too late.

In cooperation with other units of the National Research Council, the Climate Research Board expects to continue review and assessment of this important issue in order to clarify further the scientific questions involved and the range of uncertainty in the principal conclusions. We hope that this preliminary report covering but one aspect of this many-faceted issue will prove to be a constructive contribution to the formulation of national and international policies.

We are grateful to Jule Charney and to the members of the study group for agreeing to undertake this task. Their diligence, expertise, and critical judgment has yielded a report that has significantly sharpened our perception of the implications of the carbon dioxide issue and of the use of climate models in their consideration.

Verner E. Suomi, *Chairman*
Climate Research Board

PREFACE

In response to a request from the Director of the Office of Science and Technology Policy, the President of the National Academy of Sciences convened a study group under the auspices of the Climate Research Board of the National Research Council to assess the scientific basis for projection of possible future climatic changes resulting from man-made releases of carbon dioxide into the atmosphere. Specifically, our charge was

1. To identify the principal premises on which our current understanding of the question is based,

2. To assess quantitatively the adequacy and uncertainty of our knowledge of these factors and processes, and
3. To summarize in concise and objective terms our best present understanding of the carbon dioxide/climate issue for the benefit of policymakers.

The Study Group met at the NAS Summer Studies Center at Woods Hole, Massachusetts, on July 23–27, 1979, and additional consultations between various members of the group took place in subsequent weeks....

SUMMARY AND CONCLUSIONS

We have examined the principal attempts to simulate the effects of increased atmospheric CO_2 on climate. In doing so, we have limited our considerations to the direct climatic effects of steadily rising atmospheric concentrations of CO_2 and have assumed a rate of CO_2 increase that would lead to a doubling of airborne concentrations by some time in the first half of the twenty-first century. As indicated in Chapter 2 of this report, such a rate is consistent with observations of CO_2 increases in the recent past and with projections of its future sources and sinks. However, we have not examined anew the many uncertainties in these projections, such as their implicit assumptions with regard to the workings of the world economy and the role of the biosphere in the carbon cycle. These impose an uncertainty beyond that arising from our necessarily imperfect knowledge of the manifold and complex climatic system of the earth.

When it is assumed that the CO_2 content of the atmosphere is doubled and statistical thermal equilibrium is achieved, the more realistic of the modeling efforts predict a global surface warming of between 2°C and 3.5°C, with greater increases at high latitudes. This range reflects both uncertainties in physical understanding and inaccuracies arising from the need to reduce the mathematical problem to one that can be handled by even the fastest available electronic computers. It is significant, however, that none of the model calculations predicts negligible warming.

The primary effect of an increase of CO_2 is to cause more absorption of thermal radiation from the earth's surface and thus to increase the air temperature in the troposphere. A strong positive feedback mechanism is the accompanying increase of moisture, which is an even more powerful absorber of terrestrial radiation. We have examined with care all known negative feedback mechanisms, such as increase in low or middle cloud amount, and have concluded that the oversimplifications and inaccuracies in the models are not likely to have vitiated the principal conclusion that there will be appreciable warming. The known negative feedback mechanisms can reduce the warming, but they do not appear to be so strong as the positive moisture feedback. We estimate the most probable global warming for a doubling of CO_2 to be near 3°C with a probable error of ±1.5°C. Our estimate is based primarily on our review of a series of calculations with three-dimensional models of the global atmospheric circulation, which is summarized in Chapter 4. We have also reviewed simpler models that appear to contain the main physical factors. These give qualitatively similar results. . . .

To summarize, we have tried but have been unable to find any overlooked or underestimated physical effects that could reduce the currently estimated global warmings due to a doubling of atmospheric CO_2 to negligible proportions or reverse them altogether. However, we believe it quite possible that the capacity of the intermediate waters of the oceans to absorb heat could delay the estimated warming by several decades. It appears that the warming will eventually occur, and the associated regional climatic changes so important to the assessment of socioeconomic consequences may well be significant, but unfortunately the latter cannot yet be adequately projected.

NATIONAL ACADEMY OF SCIENCES

CHANGING CLIMATE

The same Office of Science and Technology Policy that requested the National Academy of Sciences assess the state of climate science in 1979 also requested in-depth follow-up assessments on carbon dioxide and climate change as the science progressed in the early 1980s. What is similar between the 1983 and 1979 reports, and what looks different? One difference comes in the expertise of the scientists involved. Unlike the 1979 assessment, William Nierenberg's Carbon Dioxide Assessment Committee included not only atmospheric scientists but also two economists, Thomas Schelling and William Nordhaus, to help flesh out future climate scenarios in terms of economic growth and energy consumption. Despite the committee's cautionary conclusions on the physical science of climate change, Nierenberg's executive summary and synthesis report presented a much more conservative set of recommendations, essentially espousing the wait-and-see approach that the Charney report had warned against. Where do the concerns of economists emerge in the text? What about this summary makes scholars characterize it as conservative?

• •

Carbon Dioxide Assessment Committee, "Changing Climate" (Washington, DC: National Academy of Sciences, 1983). Excerpt: pp. ix–x, xii, 1, 3, 4, 55, 57–60, 61, 62, 64–65. Reprinted with permission from the National Academy of Sciences, Courtesy of the National Academies Press, Washington, D.C., 2016.

FOREWORD

The Energy Security Act of 1980, while focused on the development of synthetic fuels, also called for examination of some of the environmental consequences of their development. One such consequence perceived by the Congress was the buildup of carbon dioxide (CO_2) in the atmosphere, and the National Academy of Sciences (NAS) and the Office of Science and Technology Policy (OSTP) of the Executive Office of the President were requested to prepare an assessment of its implications.

Concern about the atmosphere's carbon dioxide and its influence on climate dates back to the last century. In the 1970s, however, with recognition of a growing world population and increasing per capita use of energy, attention markedly heightened. In 1977 the National Research Council issued a report, *Energy and Climate*, prepared by a panel chaired by Roger Revelle, calling for an intensified program of research on CO_2. At around this time, the federal government began expanding its concern with CO_2, primarily through a research and assessment program in the Department of Energy. In a congressional symposium on CO_2 and energy policy in 1979 some scientists expressed the fear that atmospheric CO_2 could double by the first decade of the twenty-first century if coal and fossil-based synthetic fuels were vigorously exploited.

Such concerns and the increasing volume of research results led the Congress and the Executive to ask the NAS to consider anew various aspects of the issue. In July 1979, a brief preliminary statement about CO_2 and energy policy was released by the Academy, and later in that same summer a Panel of the Climate Research Board chaired by the late Jule Charney undertook an evaluation of the models being used to estimate likely effects of CO_2 on climate. In the following winter and spring, a committee chaired by Thomas C. Schelling and including several other members of the current Committee considered CO_2....

PREFACE

There is a broad class of problems that have no "solution" in the sense of an agreed course of action that would be expected to make the problem go away. These problems can also be so important that they should not be avoided or ignored until the fog lifts. We simply must learn to deal more effectively with their twists and turns as they unfold. We require sensible regular progress to anticipate what these developments might be with a balanced diversity of approaches. The payoff is that we will have had the chance to consider alternative courses of action with some degree of calm before we may be forced to choose among them in urgency or have them forced on us when other perhaps better options have been lost. Increasing atmospheric CO_2 and its climatic consequences constitute such a problem.

Research developments are taking place rapidly in this area. In the pages that follow we report our understanding of the status of a number of selected, critical aspects and comment on how well we think the overall attack on this complex matter is proceeding. Our stance is conservative: we believe there is reason for caution, not panic. Since understanding and proof of what is happening to climate as a result of practices that load the atmosphere with CO_2 may come too late to allow for corrective action, we may not be able to wait to make certain there is a best course. Thus, we must proceed in a manner that keeps open our major options on energy development and use, on water management, agricultural adjustment, and other relevant activities, as we move from one set of uncertainties to another. We make an effort in this report to point the way as we see it today. . . .

EXECUTIVE SUMMARY

1. Carbon dioxide (CO_2) is one of the gases of the atmosphere important in determining the Earth's climate. In the last generation the CO_2 concentration in the atmosphere has increased from 315 parts per million (ppm) by volume to over 340 ppmv.
2. The current increase is primarily attributable to burning of coal, oil, and gas; future increases will similarly be determined primarily by fossil fuel combustion. Deforestation and land use changes have probably been important factors in atmospheric CO_2 increase over the past 100 years.
3. Projections of future fossil fuel use and atmospheric concentrations of CO_2 embody large uncertainties that are to a considerable extent

irreducible. The dominant sources of uncertainty stem from our inability to predict future economic and technological developments that will determine the global demand for energy and the attractiveness of fossil fuels. We think it most likely that atmospheric CO_2 concentration will pass 600 ppm (the nominal doubling of the recent level) in the third quarter of the next century. We also estimate that there is about a 1-in-20 chance that doubling will occur before 2035....

14. The social and economic implications of even the most carefully constructed and detailed scenarios of CO_2 increase and climatic consequences are largely unpredictable. However, a number of inferences seem clear:

(a) Rapid climate change will take its place among the numerous other changes that will influence the course of society, and these other changes may largely determine whether the climatic impacts of greenhouse gases are a serious problem.

(b) As a human experience, climate change is far from novel. Large numbers of people now live in almost all climatic zones and move easily between them.

(c) Nevertheless, we are deeply concerned about environmental changes of this magnitude; man-made emissions of greenhouse gases promise to impose a warming of unusual dimensions on a global climate that is already unusually warm. We may get into trouble in ways that we have barely imagined, like release of methane from marine sediments, or not yet discovered.

(d) Climate changes, their benefits and damages, and the benefits and damages of the actions that bring them about will fall unequally on the world's people and nations. Because of real or perceived inequities, climate change could well be a divisive rather than a unifying factor in world affairs.

15. Viewed in terms of energy, global pollution, and worldwide environmental damage, the "CO_2 problem" appears intractable. Viewed as a problem of changes in local environmental factors—rainfall, river flow, sea level—the myriad of individual incremental problems take their place among the other stresses to which nations and individuals adapt. It is important to be flexible both in definition of the issue, which is

really more climate change than CO_2, and in maintaining a variety of alternative options for response....

20. With respect to specific recommendations on research, development, or use of different energy systems, the Committee offers three levels of recommendations. These are based on the general view that, if other things are equal, policy should lean away from the injection of greenhouse gases into the atmosphere.

(a) Research and development should give some priority to the enhancement of long-term energy options that are not based on combustion of fossil fuels.

(b) We do not believe, however, that the evidence at hand about CO_2-induced climate change would support steps to change current fuel-use patterns away from fossil fuels. Such steps may be necessary or desirable at some time in the future, and we should certainly think carefully about costs and benefits of such steps; but the very near future would be better spent improving our knowledge (including knowledge of energy and other processes leading to creation of greenhouse gases) than in changing fuel mix or use.

(c) It is possible that steps to control costly climate change should start with non-CO_2 greenhouse gases. While our studies focused chiefly on CO_2, fragmentary evidence suggests that non-CO_2 greenhouse gases may be as important a set of determinants as CO_2 itself. While the costs of climate change from non-CO_2 gases would be the same as those from CO_2, the control of emissions of some non-CO_2 gases may be more easily achieved....

1.3.3 The Problem of Unease about Changes of This Magnitude

... While people may be able to adapt readily to climatic change, they may be unwilling to accept climatic changes imposed on them involuntarily by the decisions of others. Thus, in trying to clarify our unease about CO_2-induced climatic change, it is necessary to point out the potentially divisive nature of the issue. It is important to recognize the distribution of incentives for, and effects of, human-induced climatic changes. Although it might be in the interest of the world economy to restrict, at some cost, the use of fossil fuels, it is probably not in the

Table 1.11 CO_2-Induced Climatic Change: Framework for Policy Choices

POSSIBLY CHANGING BACKGROUND FACTORS	POLICY CHOICES FOR RESPONSE[a]			
	1) REDUCE CO_2 PRODUCTION	2) REMOVE CO_2 FROM EFFLUENTS OR ATMOSPHERE	3) MAKE COUNTERVAILING MODIFICATIONS IN CLIMATE, WEATHER, HYDROLOGY	4) ADAPT TO INCREASING CO_2 AND CHANGING CLIMATE
Natural warming, cooling, variability			Weather Enhance precipitation, Modify, steer hurricanes and tornadoes	Environmental controls heating/cooling of buildings, area enclosures Other adaptations habitation, health, construction, transport, military
Population global, distribution: nation, climate zone, elevation (sea level), density				Migrate—internationally, intranationally
Income global average distribution				Compensate losers—intranationally, internationally
Governments				
Industrial Emissions non-CO_2 greenhouse gases particulates			Climate Change production of gases, particulates Change albedo ice, land, ocean Change cloud cover	
Energy per capita demand fossil versus nonfossil	Energy management Reduce energy use Reduce role of fossil energy Increase role of low-carbon fuels	Remove CO_2 from effluents Dispose in ocean, land Dispose of byproducts in land, ocean		
Agriculture, forestry, land use, erosion Farming and other dust	Land use Reduce rate of deforestation	Reforest Increase standing stock, fossilize trees		Change agricultural practices: cultivation, plant genetics
Agricultural emissions (N_2O, CH_4)	Preserve undisturbed carbon rich landscapes			Change demand for agricultural products, diet Direct CO_2 effects Change crop mix Alter genetics
Water supply, demand, technology, transport, conservation, exotic sources (icebergs, desalination)			Hydrology Build dams, canals Change river courses	Improve water-use efficiency

[a] Reponses may be considered at individual, local, national, and international levels.

interest of any single region or nation to incur on its own the cost of reduction in global CO_2. For example, countries that view heavy rains as disasters and countries that view them as water for their crops would have different preferences about which, if any, rains to avoid or restore and whether they or another country should forgo (or burn) fossil fuels to help effect the change. The marginal effects of climatic change on the distribution of wealth may range from quite positive to quite negative. In short, CO_2-induced climatic changes, and more generally weather and climate modification, may be a potent source of international conflict....

1.4.3 Categories of Response

Schelling (Table 1.11) develops a framework consisting of four categories of response, arrayed against background climate and trends. Category 1 is prevention, containing options for affecting the production of CO_2. Category 2 is removal: if you cannot help producing too much CO_2, can you remove some? Category 3 consists of policies deliberately intended to modify climate and weather: if too much CO_2 is produced and not enough can be removed, so that concentration is going to increase and climate is going to change in systematic fashion, can we do something about climate? Finally, Category 4 is adaptation, consisting of all the policies or actions taken in consequence of anticipated or experienced climate change....

1.4.4 Reprise

Overall, we find in the CO_2 issue reason for concern, but not panic. Although the prospect of historically unprecedented climatic changes is troubling, the problems that may be associated with it are of quite uncertain magnitude, and both climate change and increased CO_2 may also bring benefits. There are theory and evidence for each link in the chain of causal inference that we have described, but it could be that emissions will be low, or that concentrations will rise slowly, or that climatic effects will be small, or that environmental and societal impacts will be mild. Thus, we make some tentative suggestions about actual and near-term changes of policies, firmer recommendations about applied research and development with regard to the possibility of a CO_2-induced climatic change, and strong recommendations about acquiring more knowledge of various aspects of the CO_2 question. In our judgment, the knowledge we can gain in coming years should be more beneficial than a lack of action will be damaging; a program of action without a program for learning could be costly and ineffective. In the words of one reviewer of the manuscript of this report, our recommendations call for "research, monitoring, vigilance, and an open mind." ...

1.5.2 Actual and Near-Term Change of Policies

We recommend caution in undertaking any major changes in current behavior and policies solely on account of CO_2. It is probably wiser not to act aggressively to "solve the CO_2 problem" right now when we really do not know the future consequences or context of CO_2 increase. In trying to consider the world of 50 or 100 years from now, we cannot be sure that we can tell the difference between solutions and problems. It is instructive to look back at, say, 1905 to see if even the best guesses made at that time about accumulating world problems and their solutions were actually valid or useful as planning guides for the twentieth century. Life has changed since then in many unexpected ways; penicillin and air transport are vivid examples. ...

1.5.3 Energy Research and Policy

... Thus, we conclude that

1. The possibilities that concerns about the CO_2 issue will become more serious provide strong arguments for stimulating research on nonfossil energy sources. We may find that emissions are rising rapidly, that the fraction remaining airborne is high, that climate is very sensitive to CO_2 increase, or that the impacts of climate change are costly and divisive. In such a case, we want to have an enhanced ability to make a transition to nonfossil fuels.
2. The potential disruptions associated with CO_2-induced climatic change are sufficiently serious to make us lean away from fossil fuel

energy options, if other things are equal. However, our current assessment of the probability of an alarming scenario justifies primarily increased monitoring and vigilance and not immediate action to curtail fossil fuel use.

3. Analysis of prospective CO_2 emissions does not offer a strong argument for making choices among particular patterns of fossil fuel use at this time.

ENVIRONMENTAL PROTECTION AGENCY

CAN WE DELAY A GREENHOUSE WARMING?

The Effectiveness and Feasibility of Options to Slow a Build-up of Carbon Dioxide in the Atmosphere

Not long after the Nierenberg report advocated a cautious, science-first approach to the CO_2 question via the National Academy of Sciences, Stephen Seidel of the Environmental Protection Agency's Office of Policy Analysis released his own assessment of the problem of CO_2 and climate. By far the most pessimistic report released to date, Seidel and coauthor Dale Keyes found that virtually no politically feasible governmental response to rising CO_2 could significantly alter the outcome of CO_2-induced warming in the twentieth and twenty-first centuries. Seidel and Keyes asked a fundamentally different question than the Carbon Dioxide Assessment Committee—what options exist to slow CO_2-induced warming, and how effective are they?—but the report was taken as a repudiation of the conservatism of the Nierenberg report. For many in the news media and in government, however, the conflict between the reports' conclusions reflected the overall uncertainties of climate science rather than a nuanced difference in the questions asked. If the EPA and NAS couldn't agree, how certain could scientists really be about the threat of global warming? *Uncertainty*, both as a scientific concept referring to the likelihood of a "null hypothesis" and as colloquial word associated with general ignorance, served to discredit

---

Stephen Seidel, Dale L. Keyes, and U.S. Environmental Protection Agency Office of Policy and Resource Management Strategic Studies Staff, *Can We Delay a Greenhouse Warming? The Effectiveness and Feasibility of Options to Slow a Build-up of Carbon Dioxide in the Atmosphere* (1983). Excerpt: pp. i–ii, v–vii, ix.

climate scientists and privilege the wait-and-see mentality. The *New York Times* article that follows the EPA report here underscores the media's difficulty in handling the two reports together.

• •

EXECUTIVE SUMMARY

Evidence continues to accumulate that increases in atmospheric carbon dioxide (CO_2) and other "greenhouse" gases will substantially raise global temperature. While considerable uncertainty exists concerning the rate and ultimate magnitude of such a temperature rise, current estimates suggest that a 2°C (3.6°F) increase could occur by the middle of the next century, and a 5°C (9°F) increase by 2100. Such increases in the span of only a few decades represent an unprecedented rate of atmospheric warming.

Temperature increases are likely to be accompanied by dramatic changes in precipitation and storm patterns and a rise in global average sea level. As a result, agricultural conditions will be significantly altered, environmental and economic systems potentially disrupted, and political institutions stressed.

Responses to the threat of a greenhouse warming are polarized. Many have dismissed it as too speculative or too distant to be of concern. Some assume that technological options will emerge to prevent a warming or, at worst, to ameliorate harmful consequences. Others argue that only an immediate and radical change in the rate of CO_2 emissions can avert worldwide catastrophe. The risks are high in pursuing a "wait and see" attitude on one hand, or in acting impulsively on the other.

This study aims to shed light on the debate by evaluating the usefulness of various strategies for slowing or limiting a global warming. Better information is essential if scientific researchers, policymakers, and private sector decisionmakers are to work together effectively in addressing the threat of climate change....

SUMMARY OF FINDINGS

Our analysis of energy and nonenergy policies to slow or limit a global warming produced the following results:

Only One of the Energy Policies Significantly Postpones a 2°C Warming

- Worldwide taxes of up to 300% of the cost of fossil fuels (applied proportionately based on CO_2 emissions from each fuel) would delay a 2°C warming only about 5 years beyond 2040.
- Fossil fuel taxes applied to just certain countries or applied at a 100% rate would not affect the timing of a 2°C rise.
- A ban on synfuels and shale oil would delay a 2°C warming by only 5 years.
- Only a ban on coal instituted by 2000, would effectively slow the rate of temperature change and delay a 2°C change until 2055. A ban on both coal and shale oil would delay it an additional 10 years—until 2065. . . .

These findings are illustrated in the following chart. Each bar represents the number of years the 2°C date is delayed (bar above line) or advanced (bar below line), compared with the Mid-range Baseline projections.

Bans on Coal and Shale Oil Are Most Effective in Reducing Temperature Increases in 2100

- A worldwide ban on coal (and thus coal-derived synfuels) instituted by 2000 would reduce temperature change by 30% (from 5°C to 3.5°C).
- Together, a ban on shale oil and coal would reduce the projected warming in 2100 from 5°C to 2.5°C.
- Bans on shale oil alone or synfuels alone would be less effective.
- A 100% worldwide tax would reduce warming by less than 1.0°C in 2100.

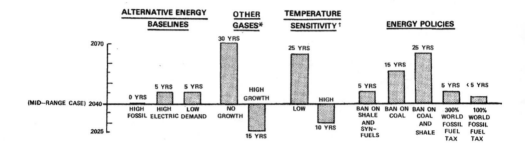

CHANGES IN THE DATE OF A 2° C WARMING

(PROJECTED DATE IN MID-RANGE BASELINE: 2040)

*REFERS TO GREENHOUSE GASES OTHER THAN CO₂: NITROUS OXIDE, METHANE, AND CHLOROFLUOROCARBONS.
†REFERS TO THE TEMPERATURE RISE IN RESPONSE TO A GIVEN INCREASE IN GREENHOUSE GASES ONCE AN EQUILIBRIUM HAS BEEN REACHED.

A Ban on Coal Seems Economically and Politically Infeasible

- Though detailed estimates of total costs of a ban on coal were beyond the scope of this study, initial approximations based only on asset losses and increases in prices of alternative fuels suggest that a coal ban is economically infeasible.
- A worldwide ban on coal also appears to be politically infeasible. Because the burden would be unevenly distributed (e.g., most of the world's coal is concentrated in only three nations, and use of coal varies dramatically between developed and developing nations), worldwide cooperation required to ban coal is unlikely.

At Best, Nonenergy Options to Limit Global Warming Are Highly Speculative

- Scrubbing CO_2 emissions from power plants is of limited effectiveness and prohibitively expensive.
- Capturing ambient CO_2 through massive forestation would place too great a burden on land, fertilizer, and irrigation requirements.

- In theory, adding SO_2 to the stratosphere might counterbalance the greenhouse warming effect, but at great cost. Moreover, the effectiveness and potential adverse environmental consequences of this proposal require much additional research. . . .

Our analysis underscores the need to reduce remaining scientific uncertainties as quickly as possible. Substantial increases in global warming may occur sooner than most of us would like to believe. In the absence of growing international consensus on this subject, it is extremely unlikely that any substantial actions to reduce CO_2 emissions could or would be taken unilaterally. Adaptive strategies undertaken by individual countries appear to be a better bet. But for these strategies to succeed, much more precise and detailed information will be needed on the timing and regionally disaggregated consequences of a global warming.

· ·

NEW YORK TIMES

HOW TO LIVE IN A GREENHOUSE

· ·

Two rather different reports on the "greenhouse effect" were released in Washington last week. One holds that the process leading to a serious warming of the earth's atmosphere is already so firmly set in motion that nothing short of a ban on coal could prevent it. The other concludes it lies so far ahead that no preventive steps are even worth attempting for the next 20 years. Which group of scientists should be believed?

The greenhouse effect is the prediction that the earth will get hotter as carbon dioxide gas released from the combustion of coal builds up in the atmosphere and traps the sun's heat like the glass of a greenhouse. Climatologists, having debated the effect for a century, now agree that

---

"How to Live in a Greenhouse," *New York Times,* October 23, 1983.

the theory is right. If carbon dioxide increases enough, ice sheets of Greenland and the west Antarctic could melt and raise sea level by up to 20 feet. Vast climate changes would disrupt agriculture and change fertile lands to desert. Boston would enjoy the climate of Miami, but most of it would lie under water.

The two new reports, one prepared for the Environmental Protection Agency and the other for the National Academy of Sciences, have more in common than might at first appear. They agree on the theory of the greenhouse effect, but differ on the numbers to be plugged into it. Perhaps because of the guidance of two economists, William Nordhaus and Thomas Schelling, the Academy's report is more tentative in predicting change and more confident of the adaptability of economic systems.

Surely the seas may rise and crop-growing belt may be pushed northward. But, says the Academy report, dikes could be built around coastal cities like Boston, just as the Dutch have walled out the sea for centuries. Crops could be genetically engineered to grow in changed climates. Indeed, all the predictable effects of a global warming can be coped with. The unpredictable effects, perhaps a sudden release of methane gas stored in ocean sediments, are the only thing we have to fear. The Academy's message, in brief, is that the greenhouse effect is for real but we can live with it.

That needed saying, because the greenhouse effect still has its skeptics. A major embarrassment for the theory is that carbon dioxide content of the atmosphere has been steadily rising for the last 25 years, yet the predicted warming has not definitely appeared. Explanations can be provided, but are inevitably ad hoc. The Academy thinks that amid the natural short- and long-term fluctuations of global temperature a warming signal is discernible, and at least not clearly absent.

"A soberness and sense of urgency should underlie our response to a greenhouse warming," concludes the Environmental Protection Agency. "There is little urgency for reductions in CO_2 emissions below an uncontrolled path before A.D. 1990," states the Academy.

Preventing the CO_2 emissions from coal plants would be extraordinarily costly. Though the Academy sees no need for immediate policy changes, the greenhouse warming is a strong, long-term argument for an energy program that is as diverse as possible, emphasizing both workable nuclear power and conservation.

R. P. TURCO, O. B. TOON, T. P. ACKERMAN,
J. B. POLLACK, AND CARL SAGAN

NUCLEAR WINTER

Global Consequences of Multiple Nuclear Explosions

In the fall of 1983, three controversial issues captured the attention of American climate scientists. First, beginning in 1981, the Reagan administration had begun to reorganize—and reduce—funding for climate change research at institutions like NASA and NCAR, and even as Al Gore brought the dangers of these reductions to light in congressional hearings (see part 3), many climate scientists scrambled to find new sources of funding for new projects. Second, the reports of the NAS Carbon Dioxide Assessment Committee and the EPA report of 1983 created controversy over the reliability of climate models and over the right policy response to climate scientists' conclusions about the global atmosphere. Finally, in the last days of 1983, a group of NASA-affiliated scientists published a report that applied the predictive capacity of computer-based numerical climate models to a new question: how would the climate respond to a nuclear exchange? Headed by Carl Sagan, host of the popular *Cosmos* television series and perhaps the most famous scientist working at the time, the group contended that ash, dust, and smoke from a nuclear exchange could reflect enough sunlight to create a catastrophic set of climatic conditions that they dubbed "nuclear winter." After working on the problem for most of the summer and fall of 1983, the TTAPS group, as they were called

---

R. P. Turco, O. B. Toon, T. P. Ackerman, J. B. Pollack, and Carl Sagan, "Nuclear Winter: Global Consequences of Multiple Nuclear Explosions," *Science* 222, no. 4630 (December 1983): 1283–92. Excerpt: p. 1283. Reprinted with permission from AAAS.

(an acronym based on their last names), published their findings in a December issue of *Science*, kicking off more than four years of controversy that helped reshape climate change politics. Here is the summary from the original TTAPS paper.

• •

Summary. The potential global atmospheric and climatic consequences of nuclear war are investigated using models previously developed to study the effects of volcanic eruptions. Although the results are necessarily imprecise, due to a wide range of possible scenarios and uncertainty in physical parameters, the most probable first-order effects are serious. Significant hemispherical attenuation of the solar radiation flux and subfreezing land temperatures may be caused by fine dust raised in high-yield nuclear surface bursts and by smoke from city and forest fires ignited by airbursts of all yields. For many simulated exchanges of several thousand megatons, in which dust and smoke are generated and encircle the earth within 1 to 2 weeks, average light levels can be reduced to a few percent of ambient and land temperatures can reach $-15°C$ to $-25°C$. The yield threshold for major optical and climatic consequences may be very low: only about 100 megatons detonated over major urban centers can create average hemispheric smoke optical depths greater than 2 for weeks and, even in summer, subfreezing land temperatures for months. In a 5000-megaton war, at northern mid-latitude sites remote from targets, radioactive fallout on time scales of days to weeks can lead to chronic mean doses of up to 50 rads from external whole-body gamma-ray exposure, with a likely equal or greater internal dose from biologically active radionuclides. Large horizontal and vertical temperature gradients caused by absorption of sunlight in smoke and dust clouds may greatly accelerate transport of particles and radioactivity from the Northern Hemisphere to the Southern Hemisphere. When combined with the prompt destruction from nuclear blast, fires, and fallout and the later enhancement of solar ultraviolet radiation due to ozone depletion, long-term exposure to cold, dark, and radioactivity could pose a serious threat to human survivors and to other species.

CARL SAGAN

NUCLEAR WAR AND CLIMATIC CATASTROPHE

Some Policy Implications

The TTAPS paper in *Science* created significant controversy within the scientific community, and it rankled some of the administrators at NASA and in the Reagan administration. But what really made nuclear winter a controversial issue was the way Sagan capitalized on popular forums to spell out the policy implications of the TTAPS findings. Here, Sagan uses the nuclear winter hypothesis to make the case for disarmament in the journal *Foreign Affairs*. To what extent does Sagan's expertise as an astrophysicist inform his assessment of nuclear winter here? How about his experience as a science popularizer? To what extent is this a piece of "scientific" writing?

. . . .

Apocalyptic predictions require, to be taken seriously, higher standards of evidence than do assertions on other matters where the stakes are not as great. Since the immediate effects of even a single thermonuclear weapon explosion are so devastating it is natural to assume—even without considering detailed mechanisms—that the more or less simultaneous explosion of ten thousand such weapons all over the Northern Hemisphere might have unpredictable and catastrophic consequences.

Carl Sagan, "Nuclear War and Climatic Catastrophe: Some Policy Implications," *Foreign Affairs* 62, no. 2 (December 1983): 257–92. Excerpt: pp. 257–59, 275, 276–77.

And yet, while it is widely accepted that a full nuclear war might mean the end of civilization at least in the Northern Hemisphere, claims that nuclear war might imply a reversion of the human population to prehistoric levels, or even the extinction of the human species, have, among some policymakers at least, been dismissed as alarmist or, worse, irrelevant. Popular works that stress this theme, such as Nevil Shute's *On the Beach*, and Jonathan Schell's *The Fate of the Earth*, have been labeled disreputable. The apocalyptic claims are rejected as unproved and unlikely, and it is judged unwise to frighten the public with doomsday talk when nuclear weapons are needed, we are told, to preserve the peace. But, as the above quotations illustrate, comparably dire warnings have been made by respectable scientists with diverse political inclinations, including many of the American and Soviet physicists who conceived, devised and constructed the world nuclear arsenals.

Part of the resistance to serious consideration of such apocalyptic pronouncements is their necessarily theoretical basis. Understanding the long-term consequences of nuclear war is not a problem amenable to experimental verification—at least not more than once. Another part of the resistance is psychological. Most people—recognizing nuclear war as a grave and terrifying prospect, and nuclear policy as immersed in technical complexities, official secrecy and bureaucratic inertia—tend to practice what psychiatrists call denial: putting the agonizing problem out of our heads, since there seems nothing we can do about it. Even policymakers must feel this temptation from time to time. But for policymakers there is another concern: if it turns out that nuclear war could end our civilization or our species, such a finding might be considered a retroactive rebuke to those responsible, actively or passively, in the past or in the present, for the global nuclear arms race.

The stakes are too high for us to permit any such factors to influence our assessment of the consequences of nuclear war. If nuclear war now seems significantly more catastrophic than has generally been believed in the military and policy communities, then serious consideration of the resulting implications is urgently called for.

It is in that spirit that this article seeks, first, to present a short summary, in lay terms, of the climatic and biological consequences of nuclear war that emerge from extensive scientific studies conducted over the past two years, the essential conclusions of which have now been endorsed by a large number of scientists. These findings were presented in detail at a special conference in Cambridge, Mass., involving almost 100 scientists on April 22–26, 1983, and were publicly announced at a conference in Washington, D.C., on October 31 and November 1, 1983. They have been reported in summary form in the press, and a detailed statement of the findings and their bases will be published in *Science*. The present summary is designed particularly for the lay reader.

Following this summary, I explore the possible strategic and policy implications of the new findings. They point to one apparently inescapable conclusion: the necessity of moving as rapidly as possible to reduce the global nuclear arsenals below levels that could conceivably cause the kind of climatic catastrophe and cascading biological devastation predicted by the new studies. Such a reduction would have to be to a small percentage of the present global strategic arsenals. . . .

The foregoing probable consequences of various nuclear war scenarios have implications for doctrine and policy. Some have argued that the difference between the deaths of several hundred million people in a nuclear war (as has been thought until recently to be a reasonable upper limit) and the death of every person on Earth (as now seems possible) is only a matter of one order of magnitude. For me, the difference is considerably greater. Restricting our attention only to those who die as a consequence of the war conceals its full impact. . . .

For me, the new results on climatic catastrophe raise the stakes of nuclear war enormously. But I recognize that there are those, including some policymakers, who feel that the increased level of fatalities has little impact on policy, but who nevertheless acknowledge that the newly emerging consequences of nuclear war may require changes in specific points of strategic doctrine. I here set down what seem to me some of the more apparent such implications, within the context of present nuclear stockpiles. The idea of a crude threshold, very roughly around 500 to 2,000 warheads, for triggering the climatic catastrophe will be central to some of these considerations. (Such a threshold applies only to something like the present distribution of yields in the strategic arsenals. Drastic conversion to very low-yield arsenals—see below—changes some of the picture dramatically.) I hope others will constructively examine these preliminary thoughts and explore additional implications of the TTAPS results....

A first strike scenario, in which the danger to the aggressor nation depends upon the unpredictable response of the attacked nation, seems risky enough. (The hope for the aggressor nation is that its retained second-strike force, including strategic submarines and unlaunched land-based missiles, will intimidate the adversary into surrender rather than provoke it into retaliation.) But the decision to launch a first strike that is tantamount to national suicide for the aggressor—*even if the attacked nation does not lift a finger to retaliate*—is a different circumstance altogether. If a first strike gains no more than a pyrrhic victory of ten days' duration before the prevailing winds carry the nuclear winter to the aggressor nation, the "attractiveness" of the first strike would seem to be diminished significantly.

S. FRED SINGER

ON A "NUCLEAR WINTER"

Scientists of many political stripes took umbrage at both the scientific findings of the TTAPS paper and the political conclusions Sagan used nuclear winter to support. Conservative physicist S. Fred Singer—who later became a vocal skeptic in debates about ozone, climate change, and acid rain while affiliated with the conservative Marshall Institute think tank—was perhaps Sagan's most vocal critic from 1983 to 1985. Singer published his criticisms in a variety of forums, including this letter in a forum in *Science* titled "On a 'Nuclear Winter'."

• •

R. P. Turco, O. B. Toon, T. P. Ackerman, J. B. Pollack, and C. Sagan (TTAPS), in their article "Nuclear winter: Global consequences of multiple nuclear explosions" (23 Dec. 1983, p. 1283), predict long-lasting subfreezing temperatures over land areas after a nuclear war (the "nuclear winter"). Their article focuses on previously neglected atmospheric radiation consequences of smoke and soot from widespread conflagrations, but does not make it sufficiently clear that changes in their assumptions and a more complete treatment can yield quite different climate scenarios. . . .

While the TTAPS results are presented in a properly qualified form, it is evident from the following article by P. R. Ehrlich *et al.* (23 Dec. 1983, p. 1293) that their results are being uncritically accepted by many. That

---

S. Fred Singer, "On a 'Nuclear Winter,'" *Science* 227, no. 4685 (January 1985): 356–58.

is not to say that the long-term consequences of a nuclear exchange should be discounted. Even with the TTAPS predictions reversed, a hot earth surface could threaten the survival of animals and plants. But then again, the temperature change might be negligible and so would the biological consequences.

The same issue of *Science* contains a News and Comment briefing (p. 1308) about a joint American-Soviet scientific forum sponsored by the Nuclear Freeze Foundation on 8 December 1983. This forum sharply criticized a study prepared by the Federal Emergency Management Agency (FEMA), which suggests that food supplies would still be available after a nuclear attack. The FEMA study was faulted not only for its conclusions, "but also for its *underlying attitudes*" (emphasis mine). Senator Edward Kennedy (D-Mass.) was quoted as saying: "This kind of thinking makes nuclear war more likely because it makes nuclear war seem more bearable."

This remark raises ethical problems. First, does prediction of a global holocaust make nuclear war less likely? And, second, should scientists therefore ignore scenarios which produce less severe global outcomes?

S. Fred Singer
George Mason University
Fairfax, Virginia 22030

STARLEY L. THOMPSON AND STEPHEN H. SCHNEIDER

NUCLEAR WINTER REAPPRAISED

If Singer launched attacks on Sagan's scientific conclusions and political recommendations from the right, by 1986 a number of Sagan's political allies with left-leaning viewpoints had also begun to reassess the basic science of TTAPS. In particular, Sagan's friend Stephen Schneider and his NCAR colleague Starley Thompson revisited the TTAPS model and then followed Sagan into the pages of *Foreign Affairs* with a set of conclusions that in some ways undermined the TTAPS paper. The episode—which also involved personal correspondences and meetings, public presentations, and other media outlets—drove a lasting wedge between Schneider and Sagan despite a wealth of common scientific and political ground.

• •

Apocalyptic visions of the environmental effects of nuclear war have been a part of our popular culture for decades. But apart from appreciating any entertainment value, the cognoscenti of nuclear war have regarded the doomsday predictions as ignorant at best, or dangerous propaganda at worst. The potential global environmental effects of nuclear explosions that were known before 1982—radioactive fallout and the destruction of the stratospheric ozone layer—were almost universally accepted in the strategic weapons community as being far short of true doomsday proportions. Indeed, for the combatant nations, such uncertain "secondary" effects were thought to pale before the assured

---

Starley L. Thompson and Stephen H. Schneider, "Nuclear Winter Reappraised," *Foreign Affairs* 64, no. 5 (July 1986): 981–1005. Excerpt: pp. 981–84.

direct effects of blast, heat and local radioactivity. From a scientific standpoint, this skepticism of environmental doomsday effects was probably justified in the sense that a large nuclear war would have been more devastating to the superpowers than any known indirect effects. The discovery of "nuclear winter" has challenged this skepticism because it has been much more compelling scientifically than the earlier predictions of global environmental effects. It has even been referred to as an inadvertent manifestation of Herman Kahn's "doomsday machine."

The nuclear winter hypothesis, stated simply, contended that the smoke and dust placed in the atmosphere by a large nuclear war would prevent most sunlight from reaching the earth's surface and produce a widespread cooling of land areas. The first two climatic conclusions of the theory were the most important: effects would be severe (weeks of sub-freezing temperatures), and effects would be widespread (at least hemispheric in scale). These grim scientific conclusions gave rise to two unique implications: the possibility of human extinction, and the potential suicide of an attacker even without retaliation by the attacked party. These implications, if confirmed, would indeed approach the definition of the traditional doomsday machine.

Another assertion was added to the hypothesis in the form of a scientific judgment: namely, that a "threshold" existed above which the climatic effects of a nuclear attack would become catastrophic. Thus, this doomsday machine did not possess a hair trigger, and would allow nuclear wars to be fought at some level substantially below the destructive potential of the current nuclear arsenals without global climatic catastrophe.

An additional major scientific conclusion closely followed the announcements of severe and widespread effects, but it initially received much less attention. Despite early suggestions that a nuclear winter was quite probable as long as a substantial number of large cities were attacked, many scientists concluded that the magnitude of effects would indeed be strongly dependent on uncertain, or even unknowable, factors.

The severe conclusions about nuclear winter provoked a broad spectrum of suggested responses for strategic policy. For many who took nuclear winter seriously, a perceived solution was a drastic reduction of nuclear arms to a level no greater than that necessary to constitute a minimal deterrent. On the other hand, the U.S. Department of Defense—which accepted the possibility of nuclear winter—argued that strengthening deterrence, combined with more research into nuclear winter, would be the best response. Thus, the potential for nuclear winter effects was used by the Department of Defense as another reason to support the continued modernization of U.S. strategic forces with smaller, more accurate warheads, and to pursue the the ongoing modernization of strategic doctrine—expanded kinds of options for nuclear strikes and options for "limited" strikes—would reduce the probability of nuclear war by providing a more credible deterrent.

Cynics and agnostics were also well represented in the policy debate. Many considered the conclusions drawn from the nuclear winter theory, especially the early ones, to be too uncertain to provide the basis for any discussion of policy implications. Others pointed out that the horrible effects of nuclear war for combatant nations were already widely accepted, that any environmental or indirect effects on noncombatants would not further motivate the superpowers to seek arms control, and that additional bad effects would only add marginally to an already strong mutual deterrence.

We intend to show that on scientific grounds the global apocalyptic conclusions of the initial nuclear winter hypothesis can now be relegated to a vanishingly low level of probability. Thus the argument that nuclear winter provides the sole basis for drastic strategic arms reductions has been greatly weakened. But, at the same time, there is little that is thoroughly understood about the environmental effects of a nuclear war. In particular, we do not think that all environmental effects should once again be considered as "secondary." Important environmental and widespread societal effects of nuclear war remain quite probable and do suggest further scientific and policy considerations. Our current understanding of environmental effects will be reviewed and then used to bolster arguments for strengthened strategic stability, not necessarily excluding newer strategic systems, but at significantly reduced levels of arsenals.

JAMES HANSEN

TESTIMONY BEFORE THE SENATE COMMITTEE ON ENERGY AND NATURAL RESOURCES, JUNE 23, 1988

In the summer of 1988—just two years after the last salvos of the nuclear winter battles—climate scientists began to realize that 1987 and 1988 were shaping up to be two of the warmest years ever recorded throughout the globe. Heat waves and droughts in the eastern United States, India, and Russia rekindled many of the climate-related food concerns articulated in the mid-1970s, and global warming began to emerge more frequently and more sensationally in the national media. In June 1988, Colorado senator Tim Wirth and NASA scientists James Hansen and Suki Manabe capitalized on the heat to make a firm statement about climate change that would become one of the most famous catalytic incidents in the history of global warming politics. On a sizzling day in Washington, DC, in late June with nothing else newsworthy on the congressional calendar, Wirth held a second set of hearings on climate change before the Senate Committee on Energy and Natural Resources and invited Hansen to testify as an expert witness. In his testimony, Hansen argued unequivocally that global warming was real and that he could say with 99 percent certainty that its impacts were already being felt in 1988. The 101° heat wasn't necessarily the result of global warming, Hansen explained, but in a warming world Americans could expect a lot more 101° days like that one. For scientists, Hansen's claims that his models had detected global warming's impacts—and his claim of 99 percent certainty—stirred significant controversy. The public message

---

Greenhouse Effect and Global Climate Change: Hearings of the Senate Committee on Energy and Natural Resources, 100th Cong. (1988). Excerpt: pp. 39–40.

Hansen's testimony sent was straightforward and effective, however. After the hearing, Hansen famously told *New York Times* reporter Philip Shabecoff, "It is time to stop waffling so much and say that the evidence is pretty strong that the greenhouse effect is here."*

• •

Dr. HANSEN. Mr. Chairman and committee members, thank you for the opportunity to present the result of my research on the greenhouse effect which has been carried out with my colleagues at the NASA Goddard Institute for Space Studies.

I would like to draw three main conclusions. Number one, the earth is warmer in 1988 than at any time in the history of instrumental measurements. Number two, the global warming is now large enough that we can ascribe with a high degree of confidence a cause and effect relationship to the greenhouse effect. And number three, our computer climate simulations indicate that the greenhouse effect is already large enough to begin to affect the probability of extreme events such as summer heat waves.

My first viewgraph, which I would like to ask Suki to put up if he would, shows the global temperature over the period of instrumental records which is about 100 years. The present temperature is the highest in the period of record. The rate of warming in the past 25 years, as you can see on the right, is the highest on record. The four warmest years, as the Senator mentioned, have all been in the 1980s. And 1988 so far is so much warmer than 1987, that barring a remarkable and improbable cooling, 1988 will be the warmest year on the record.

Now let me turn to my second point which is causal association of the greenhouse effect and the global warming. Causal association requires first that the warming be larger than natural climate variability and, second, that the magnitude and nature of the warming be consistent with the greenhouse mechanism. These points are both addressed in my second viewgraph. The observed warming during the past 30 years, which is the period when we have accurate measurements of atmospheric

---

* Philip Shabecoff, "Sharp Cut in Burning of Fossil Fuels Is Urged to Battle Shift in Climate," *New York Times*, June 24, 1988.

composition, is shown by the heavy black line in this graph. The warming is almost 0.4 degrees Centigrade by 1987 relative to climatology, which is defined as the 30 year mean, 1950 to 1980 and, in fact, the warming is more than 0.4 degrees Centigrade in 1988. The probability of a chance warming of that magnitude is about 1 percent. So, with 99 percent confidence we can state that the warming during this time period is a real warming trend....

Then my third point. Finally, I would like to address the question of whether the greenhouse effect is already large enough to affect the probability of extreme events, such as summer heat waves. As shown in my next viewgraph, we have used the temperature changes computed in our global climate model to estimate the impact of the greenhouse effect on the frequency of hot summers in Washington, D.C., and Omaha, Nebraska. A hot summer is defined as the hottest one-third of the summers in the 1950–1980 period, which is the period the Weather Bureau uses for defining climatology. So, in that period the probability of having a hot summer was 33 percent, but by the 1990s, you can see that the greenhouse effect has increased the probability of a hot summer to somewhere between 55 and 70 percent in Washington according to our climate model simulations. In the late 1980s, the probability of a hot summer would be somewhat less than that....

I believe that this change in the frequency of hot summers is large enough to be noticeable to the average person. So, we have already reached a point that the greenhouse effect is important. It may also have important implications other than for creature comfort.

HISTORICIZING DATA

Data images make odd cultural artifacts. On one hand, scientists present their data in images as a form of visual communication, intended, like other forms of visual culture, to convey both specific information and larger culturally coded messages. On the other hand, however, scientists typically hew to methods of measurement and mathematical analysis intended to ensure that the data they present reflects some objective reality that transcends the cultural. Ultimately the data is supposed to "speak for itself."

Such is the case with the Keeling Curve, the oscillating, upward-sloping graph of measured atmospheric CO_2 that has come to stand as one of the most important and powerful scientific symbols of anthropogenic climate change. To a lay reader, it may seem odd to read a simple measure of atmospheric gas through the many-sided prism of modern American life the way you might read a historical photograph or piece of art. And yet, the Keeling Curve functions as much as a symbol in our collective cultural understanding of climate change as it does a representation of data about CO_2. The Keeling Curve faithfully represents something quite real—the accumulation of CO_2 in the atmosphere since 1958, expressed in parts per million (ppm)—but it is also a constructed image ripe for reading, similar to a painting, a photograph, a landscape, or a written document.

The Keeling Curve is one of many data images both scientists and advocates have used to highlight the reality and the real danger of anthropogenic climate change. Like the Keeling Curve, each of these

---

A version of this essay appeared as "This Is Nature; This Is Un-nature: Reading the Keeling Curve," *Environmental History* 20, no. 2 (April 2105). Reproduced with permission.

images at once reflects a physical reality and a set of contextually specific concerns that shape the way its creators present that reality in lines, bars, and numbers. Where climate change is concerned, many of these data images portray a history of atmospheric or terrestrial change, and it is important to understand how that change has been monitored and made into an image. Equally important for the historian, however, is that the way data images portray change reflects the historical context in which those images were created. The data of the image may be historical, but the data image itself also has a history.

THE KEELING CURVE

At its heart, the Keeling Curve is a scientific image with scientific objectives, and any reading of it as a cultural symbol necessarily starts with

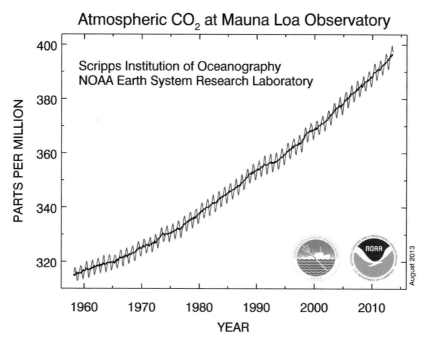

FIGURE HD.1. Keeling Curve: Atmospheric CO_2 at Mauna Loa Observatory. From Dr. Pieter Tans, NOAA/ESR; and Dr. Ralph Keeling, Scripps Institution of Oceanography, "Trends in Atmospheric Carbon Dioxide," 2014. Available at National Oceanic and Atmospheric Administration, www.esrl.noaa.gov/gmd/ccgg/trends.

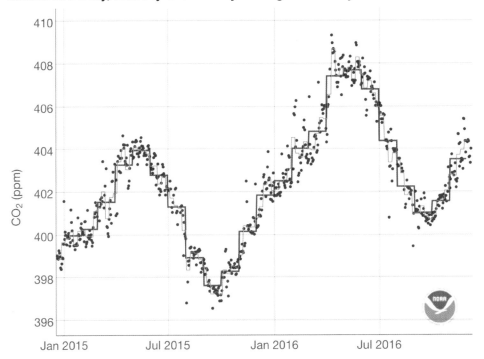

FIGURE HD.2. Keeling Curve: Mauna Loa daily, monthly and weekly averages for two years. From Dr. Pieter Tans, NOAA/ESR: and Dr. Ralph Keeling, Scripps Institution of Oceanography, 2014. Available at National Oceanic and Atmospheric Administration, www.esrl.noaa.gov/gmd/ccgg/trends/graph.html.

the scientific construction of the image. In figure HD.1, for example, the curve measures a single variable, CO_2, over time. But the graph is not a plot of raw data. Rather, the oscillating, upward-sloping line from 1958 to 2016 runs through points representing the average monthly mean of the daily averages of measured atmospheric CO_2 at the Mauna Loa Observatory, as corroborated by measurements at other observing stations around the world. Figure HD.2, an interactive image of the Keeling Curve from the National Oceanic and Atmospheric Administration's (NOAA) Earth Systems Research Laboratory, helps to demonstrate this construction process. In the image, individual measurements, aggregated and averaged over increasing increments of time, constitute recognizable trends. In figure HD.1—the most recognizable version of the curve—NOAA portrays these trends in lines that smooth out variations and irregularities in individual measurements and short-term averages—that is, outliers among the dots—in order to make meaning out of changes in CO_2 over months and years.

In most versions of the Keeling Curve, this aggregating and averaging stops at the monthly mean; we recognize the Keeling Curve not just by its upward slope but also by its peculiar wavelike shape. Frequently, depictions of the curve like the one in figure HD.1, from NOAA's Keeling Curve website, include both the oscillating line of monthly means (in red) and a straighter, more direct line of seasonally corrected data that roughly approximates the annual average measured CO_2 over time (in black). Rarely, however, do we see this straight line alone; if we do, we do not recognize it as the Keeling Curve.

In a few important respects, we need both lines to interpret the curve. The Keeling Curve of figure HD.1 is a form of storytelling, and it actually tells two mutually instructive stories. The line of annual average tells a long-term story of change, depicting a troublingly persistent rise in atmospheric CO_2 over the past half century that most scientists recognize as a consequence of humans' collective burning of fossil fuels. Embedded within that long story, however, is an important story of annual oscillation operating independently of human influence. Across the globe, plants fix atmospheric carbon in the form of new shoots, leaves, blossoms, and flowers during the spring and summer months—the growing season—and then release that carbon back into the atmosphere in the fall and winter months as they die or drop their leaves. Because the Northern Hemisphere contains the majority of the earth's continental land masses—and a corresponding majority of its seasonal plants—measurements of atmospheric CO_2 reflect the annual uptake and release of CO_2 as an artifact of the Northern Hemisphere's seasonal cycles. The annual oscillations depicted in the Keeling Curve reflect this cycle of growth and decay. The Keeling Curve thus also uses scientific data to tell the story of the earth "breathing."[1]

---

1. See Scripps Institution of Oceanography, "The Keeling Curve: A Daily Record of Atmospheric Carbon Dioxide from the Scripps Institution of Oceanography at UC San Diego," http://keelingcurve.ucsd.edu.

There are two good reasons to tell these stories together. First, the two stories—one of cyclical planetary respiration and the other of secular atmospheric change—reflect the two major insights of the curve's namesake, Charles David Keeling. In 1956, Keeling was a postdoctoral fellow in geochemistry at California Institute of Technology with Harrison Brown, who encouraged Keeling to begin investigating the transfer of carbon among water, rock, and air over time. In taking control samples for the highly sensitive gas manometer that he had designed for the carbon project, Keeling noted a consistency in the "background" concentrations of CO_2 that he took from places as different and disparate as the lab in Pasadena and the forests of the Olympic Peninsula in Washington state.[2] In 1957, Roger Revelle of the Scripps Institute of Oceanography and Harry Wexler of the U.S. Weather Bureau secured funding for Keeling to conduct continuous CO_2 measurements at the Mauna Loa Observatory, and by 1960 Keeling had both identified the cyclical variations of atmospheric CO_2 as an artifact of seasonal growth and decay and noted a consistent year-to-year increase of about 1 ppm in background measures of atmospheric CO_2.[3] To portray the two phenomena together consolidates Keeling's career into a form of well-deserved data-driven scientific hagiography.

There is a second, more important way to read these two phenomena together, however. The upward-sloping oscillations of the Keeling Curve embed a cyclical story about the processes of nature—seasonal growth and decay—within a larger story about the "unnature" of anthropogenic climate change.

In figure HD.1, the red line of monthly means oscillating up and down within the curve represents the earth breathing. This is nature. The cycle is annual, but the organic process is timeless—influenced by human activities, perhaps, but ultimately independent of anything short of tectonic change. And yet, the Keeling Curve as a whole is not timeless. The oscillating curve of monthly means does not run straight

---

2. Scripps Institution of Oceanography, "Keeling Curve History," http://keelingcurve.ucsd.edu/the-history-of-the-keeling-curve.

3. See also Spencer Weart, *The Discovery of Global Warming* (Cambridge, MA: Harvard University Press, 2003), 36; Gale E. Christiansen, *Greeenhouse: The 200 Year Story of Global Warming* (New York: Penguin, 1999), 151–57.

across the page; it slopes upward and to the right along the black line of corrected averages as it follows the half-century human time scale demarcated by the numbers at the bottom of the graph. The earth continues to inhale and exhale as it has for millions of years, but humans have changed the air it breathes. This is un-nature, or nature that humans have tipped on its side. Scientists don't have to make an explicit normative judgment about rising CO_2 when they present the two lines of the curve together. The Keeling Curve puts rising anthropogenic CO_2 in tension with a timeless natural process of planetary respiration, and extrascientific cultural ideas about the duality of the natural and the human do the work to make that image meaningful.

The tension between the natural and the unnatural embedded in the two lines of the Keeling Curve has informed the recent emphasis in climate change discourse on the concept of the anthropocene. First popularized by Paul Crutzen and E. F. Stoermer (see part 6), the term *anthropocene* describes the current geological epoch, in which humans have begun to act as agents of geophysical change, particularly in the lithosphere.[4] The core insight behind the term is that through our intensive burning of fossil fuels, humans have collectively affected geophysical changes on human timescales.[5] Few images convey this concept more clearly and effectively than the simple upward-sloping oscillations of the Keeling Curve. Cycles of planetary respiration—the single up-and-down annual units of the red line of monthly means on the curve—plotted in human time (along the straighter black line running from lower left to upper right) reveal secular planetary change.

In fact, understood in the context of a world in which human activity has collapsed the distinction between the human and the natural (however artificial that distinction may have been), the Keeling Curve becomes a new form of historical time marker. As Dipesh Chakrabarty writes, the idea of the anthropocene requires that we reevaluate narratives of modernity, especially insofar as climate change complicates the benefits of a "modernity" achieved through the burning of fossil

---

4. See Dipesh Chakrabarty, "The Climate of History: Four Theses," *Critical Inquiry* (Winter 2009): 197–222.

5. Ibid., 207–8.

fuels.[6] The Keeling Curve offers an anthropocenic timeline, a new chronological background for contemplating both personal and collective histories. To say that Lee Harvey Oswald shot President Kennedy at 317 ppm, near the lower left of figure HD.1; that Nixon launched the Cambodian incursion somewhere around 327 ppm; that Reagan fired secretary of the interior James Watt at 341.5 ppm, near the middle of the curve; and that the United States invaded Iraq once at 353 ppm and again at 376.5 ppm, closer to the top right, is to trace historical events onto a geophysical timeline that gives those events new historical and environmental meaning. Where were you at 400 ppm? Will you be around at 450 ppm?

In tracking history in anthropocenic time via measurements of CO_2, we also track a history of scientific knowledge that makes CO_2 meaningful. Much of that knowledge is not new; as this collection reveals, scientists have understood anthropogenic climate change as an important environmental and human problem for at least five decades, and the international political community has officially recognized the importance of climate change for more than twenty years through hybrid science-to-policy mechanisms like the Intergovernmental Panel on Climate Change and political mechanisms like the United Nations Framework Convention on Climate Change. In fact, Keeling himself had to know enough to look for a change in background atmospheric CO_2 to begin his measurements, and in that sense the Keeling Curve is an artifact of scientific knowledge from the 1950s as much as it is a driver of knowledge about climate change in the twenty-first century.

There have been short-term downturns in public and scientific interest in CO_2, of course, and at least one of these downturns is inscribed in a 1971 image of the curve itself. A gap in the oscillating line of monthly means near the middle of figure HD.3, which comes from the Study of Man's Impact on Climate, reflects a three-month period in

---

6. The gap was short-lived, and CO_2 would land on President Johnson's desk in a report on the state of the environment just a year later. U.S. President's Science Advisory Committee Environmental Pollution Panel, *Restoring the Quality of Our Environment: Report of the Pollution Panel, President's Science Advisory Committee* (Washington, DC: U.S. Government Printing Office, 1965).

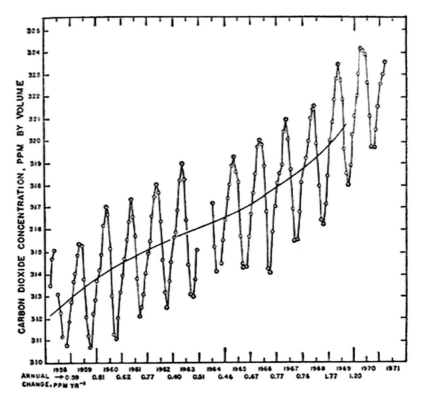

FIGURE HD.3. Keeling Curve (1971). Mean monthly values of CO_2 concentration at Mauna Loa, Hawaii, for the period 1958–1971. From William H. Matthews, William H. Kellogg, and G. D. Robinson, eds., *Man's Impact on the Climate*, figure 8.11, p. 234, © 1971 Massachusetts Institute of Technology, by permission of the MIT Press.

the spring of 1964 when Keeling lost his funding. The gap functions at once as a historical time marker of budgetary problems in 1963 and a rhetorical argument for continued funding—avoiding future gaps—in 1971. Here, in a very real way, an image of the Keeling Curve depicts not only measured atmospheric CO_2, but also the scientific politics of its measurement at a particular time in history.[7]

7. See Dipesh Chakrabarty, "The Climate of History: Four Theses," *Critical Inquiry* (Winter 2009): 197–222.

THE HOCKEY STICK

One of the most controversial data images informing climate change discourse in the last twenty years has been the so-called hockey-stick image, representing the reconstructed global mean temperature over the past thousand years. Originally introduced by Michael Mann and Raymond Bradley, then at the University of Massachusetts-Amherst, and Malcolm Hughes at the University of Arizona, the graph was called the "hockey stick" because of the relationship between the relatively flat, gently downward-sloping line of historical temperatures and the abrupt upward slope of recent warming. Controversy among scientists surrounding the hockey stick arose in part over the statistical methods that Mann, Bradley, and Hughes used to incorporate proxy data and uncertainty into the graph, which dovetailed with political controversies over climate change when the IPCC, and later Al Gore and others, used the hockey stick as one piece of evidence to show that the world was in fact warming and that the 1990s were the warmest decade in the past thousand years. As the key in figure HD.5 reveals, the noisy black line running from AD 1000 to AD 1980 represents reconstructed data, smoothed out to connect forty-year averages in the purple line running through the whole graph. The yellow shading represents the statistically defined uncertainty range for the reconstructed data—an artifact of efforts stepped up after the release of IPCC's First Assessment Report to quantify uncertainties in climate science. The rapidly upward-sloping red line represents raw temperature data gathered since 1902, while the overlap from 1902 to 1980 provides a "calibration period" to test historical temperature reconstructions.[8] Mann and his colleagues built in part off schematic reconstructions of temperature, CO_2, and methane included in IPCC 1990 (see figure HD.4). In 2001, the IPCC's Third Assessment Report drew on the hockey stick to create a series of images of past global temperatures and CO_2 levels relative to potential

---

8. The hockey stick is featured prominently in Al Gore's *An Inconvenient Truth*, with Gore using correlations between the original temperature reconstructions and historical atmospheric CO_2 content to draw an inference about future temperatures by using a cherry-picker to follow the giant projected curve up the blade of the hockey stick and off his screen.

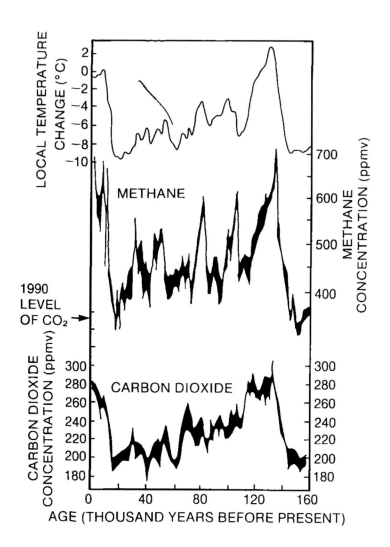

Figure 2: Analysis of air trapped in Antarctic ice cores shows that methane and carbon dioxide concentrations were closely correlated with the local temperature over the last 160,000 years. Present day concentrations of carbon dioxide are indicated

FIGURE HD.4. Intergovernmental Panel on Climate Change (1990), Temperature, Atmospheric Methane, and Atmospheric Carbon Dioxide over the past 160,000 years.

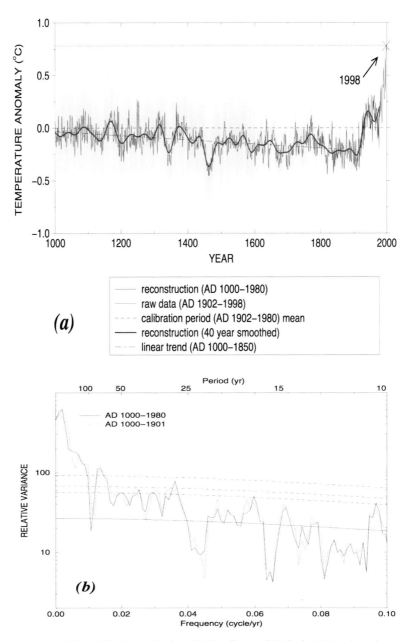

FIGURE HD.5. Michael E. Mann, Raybond S. Bradley, and Malcolm K. Hughes, the Original "Hockey Stick Graph." From "Northern Hemisphere Temperatures during the Past Millennium: Inferences, Uncertainties, and Limitations," *Geophysical Research Letters* 26, no. 6 (1999).

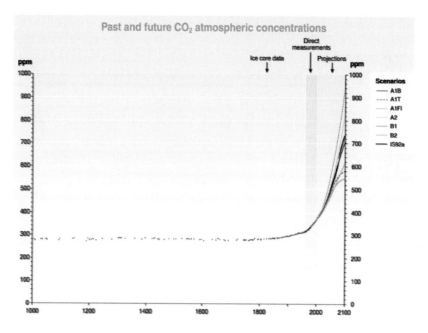

FIGURE HD.6. Intergovernmental Panel on Climate Change (2001), Figure SPM-10a: Past and Future CO2 Atmospheric Concentrations. From J. T. Houghton, G. J. Jenkins, and J. J. Ephraums, eds., *Climate Change: The IPCC Scientific Assessment (1990)* (Cambridge: Cambridge University Press, 1990), xv.

future temperature trends under a variety of emissions scenarios introduced in a 2000 IPCC supplement, the "Special Report on Emissions Scenarios," or SRES. Figure HD.6 provides a "hockey stick" of CO_2; figure HD.7 shows temperature. Mann—and the hockey stick—would later sit at the center of a very public controversy over the image when climate change skeptics accused him and his colleagues of manipulating data, a charge they supported with e-mails hacked from a server at the University of East Anglia. Mann was exonerated, but "climategate" ushered in a new round of denial in the politics of climate change.[9]

9. Michael Mann wrote a memoir about the incident in 2012 called *The Hockey Stick and the Climate Wars: Dispatches from the Front Lines* (New York: Columbia University Press, 2012).

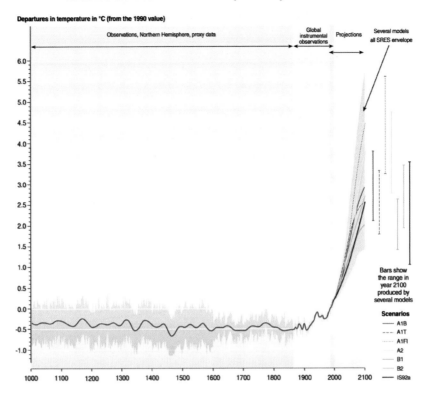

FIGURE HD.7. Intergovernmental Panel on Climate Change (2001), Figure SPM-10b: Variations of the Earth's surface temperature: years 1000–2100. From IPCC TAR SYR, *Climate Change 2001: Synthesis Report*, Contributions of Working Groups I, II, and III to the Third Assessment Report of the Intergovernmental Panel on Climate Change, 2001. Available at www.grida.no/publications/other/ipcc_tar/.

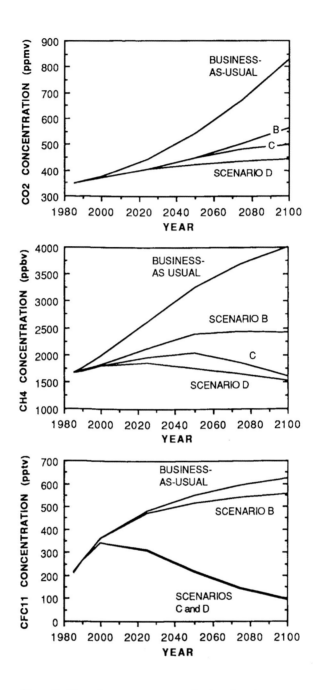

Figure 5: Atmospheric concentrations of carbon dioxide methane and CFC-11 resulting from the four IPCC emissions scenarios

SCENARIO GRAPHS

More than either of the first two IPCC assessment reports, the Third Assessment Report in 2001 self-consciously used images of climate data to tell stories about the climatic past and the potential climatic future. In particular, TAR framed the science of climate change in terms of future emissions choices by focusing on "emissions scenarios" developed by a variety of competing teams of scientists. The scenario idea was not a new one in 2001—scenario studies informed the first IPCC report—but the sophisticated integration of policy options and scientific outcomes clarified the Third Assessment Report's political message despite its scientific authorship and focus. Here, you can trace images of scenario studies as they gain in sophistication and political focus from IPCC 1990 to IPCC 2001. IPCC 1990 (figure HD.8) defined four potential emissions scenarios, including (a) a continuation of global emissions of greenhouse gases at 1990 levels; (b) a 50 percent reduction in emissions from 1990 levels; (c) a 2 percent per year reduction in emissions; and (d) a 2 percent annual increase from 1990 to 2010, followed by a 2 percent annual decrease from 2010 forward. Shortly after the release of the First Assessment Report, however, the IPCC decided to amend its scenarios to try to account for a wider array of assumptions about how greenhouse gas emissions might change even in the absence of climate change policy. The result was an IPCC supplement published in 1992 that outlined six potential emissions scenarios known as IS92(a–f), included in the Second Assessment Report (see figure HD.9). The IS92 scenarios in turn gave rise to the Special Report on Emissions Scenarios in 2000, which informed the Third Assessment Report in 2001 (see figure HD.10). The assessment reports and supplements themselves provide details on the technical differences among the emissions scenarios, but even without those technical details, the presentation of those scenarios reveals a changing relationship between the IPCC and the policy it is meant to

FIGURE HD.8.(*opposite*) Intergovernmental Panel on Climate Change (1990), Figure 5: Atmospheric concentrations of carbon dioxide, methane, and CFC-11 resulting from the four IPCC emissions scenarios. From J. T. Houghton, G. J. Jenkins, and J. J. Ephraums, eds., *Climate Change: The IPCC Scientific Assessment* (1990) (Cambridge: Cambridge University Press, 1990), xvii.

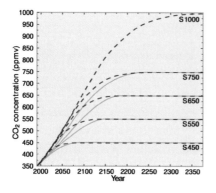

Figure 1 (a). Carbon dioxide concentration profiles leading to stabilization at 450, 550, 650 and 750 ppmv following the pathways defined in IPCC (1994) (solid curves) and for pathways that allow emissions to follow IS92a until at least the year 2000 (dashed curves). A single profile that stabilizes at a carbon dioxide concentration of 1000 ppmv and follows IS92a emissions until at least the year 2000 has also been defined. Stabilization at concentrations of 450, 650 and 1000 ppmv would lead to equilibrium temperature increases relative to 1990[14] due to carbon dioxide alone (i.e., not including effects of other greenhouse gases (GHGs) and aerosols) of about 1°C (range: 0.5 to 1.5°C), 2°C (range: 1.5 to 4°C) and 3.5°C (range: 2 to 7°C), respectively. A doubling of the pre-industrial carbon dioxide concentration of 280 ppmv would lead to a concentration of 560 ppmv and doubling of the current concentration of 358 ppmv would lead to a concentration of about 720 ppmv.

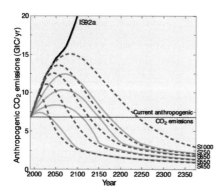

Figure 1 (b). Carbon dioxide emissions leading to stabilization at concentrations of 450, 550, 650, 750 and 1000 ppmv following the profiles shown in (a) from a mid-range carbon cycle model. Results from other models could differ from those presented here by up to approximately ± 15%. For comparison, the carbon dioxide emissions for IS92a and current emissions (fine solid line) are also shown.

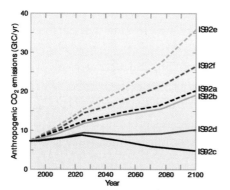

Figure 2. Annual anthropogenic carbon dioxide emissions under the IS92 emission scenarios (see Table 1 in the Summary for Policymakers of IPCC Working Group II for further details).

[14] These numbers do not take into account the increase in temperature (0.1 to 0.7°C) which would occur after 1990 because of CO_2 emissions prior to 1990.

FIGURE HD.9. Intergovernmental Panel on Climate Change (1996). Scenario Graphs. From J. Leggett, W. J. Pepper, R. J. Swart, J. Edmonds, L. G. Meira Filho, I. Mintzer, M. X. Wang, and J. Watson, "Emissions Scenarios for the IPCC: an Update," in *Climate Change 1992: The Supplementary Report to the IPCC Scientific Assessment* (Cambridge: Cambridge University Press, 1992).

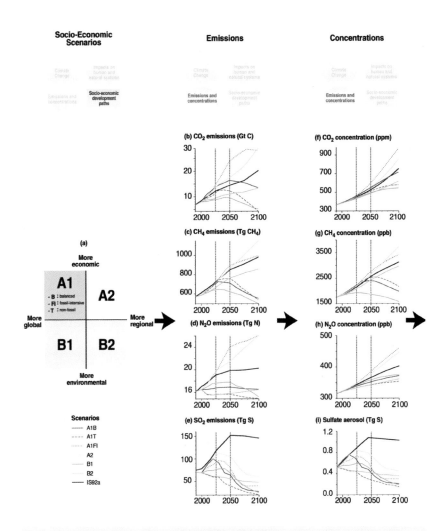

FIGURE HD.10. Intergovernmental Panel on Climate Change (2001). Figure SPM-3: Socio-Economic Scenarios. From IPCC TAR SYR, *Climate Change 2001: Synthesis Report*, Contributions of Working Groups I, II, and III to the Third Assessment Report of the Intergovernmental Panel on Climate Change, 2001, www.grida.no/publications/other/ipcc_tar/.

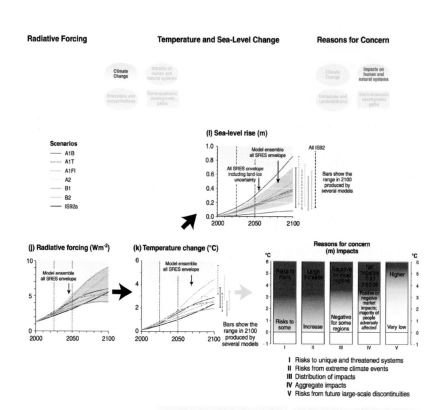

FIGURE HD.11. "Burning Embers." Intergovernmental Panel on Climate Change (2001). Figure SPM-3: Socio-Economic Scenarios. From IPCC TAR SYR, *Climate Change 2001: Synthesis Report*, Contributions of Working Groups I, II, and III to the Third Assessment Report of the Intergovernmental Panel on Climate Change, 2001, www.grida.no/publications/other/ipcc_tar/.

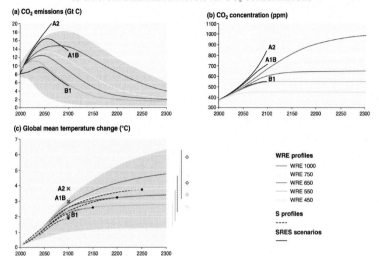

Figure SPM-6: Stabilizing CO₂ concentrations would require substantial reductions of emissions below current levels and would slow the rate of warming.
a) CO_2 emissions: The time paths of CO_2 emissions that would lead to stabilization of the concentration of CO_2 in the atmosphere at various levels are estimated for the WRE stabilization profiles using carbon cycle models. The shaded area illustrates the range of uncertainty.
b) CO_2 concentrations: The CO_2 concentrations specified for the WRE profiles are shown.
c) Global mean temperature changes: Temperature changes are estimated using a simple climate model for the WRE stabilization profiles. Warming continues after the time at which the CO_2 concentration is stabilized (indicated by black spots), but at a much diminished rate. It is assumed that emissions of gases other than CO_2 follow the SRES A1B projection until the year 2100 and are constant thereafter. This scenario was chosen as it is in the middle of the range of SRES scenarios. The dashed lines show the temperature changes projected for the S profiles (not shown in panels (a) or (b)). The shaded area illustrates the effect of a range of climate sensitivity across the five stabilization cases. The colored bars on the righthand side show uncertainty for each stabilization case at the year 2300. The diamonds on the righthand side show the average equilibrium (very long-term) warming for each CO_2 stabilization level. Also shown for comparison are CO_2 emissions, concentrations, and temperature changes for three of the SRES scenarios.

FIGURE HD.12. Intergovernmental Panel on Climate Change (2001). Figure SPM-6: CO₂ Stabilization Scenarios. From IPCC TAR SYR, *Climate Change 2001: Synthesis Report*, Contributions of Working Groups I, II, and III to the Third Assessment Report of the Intergovernmental Panel on Climate Change, 2001, www.grida.no /publications/other/ipcc_tar/.

underpin. The embedded image of "Reasons for Concern: (m) Impacts" deserves special attention here. The image, commonly referred to as "Burning Embers" (figure HD.11), became a subject of controversy when it was removed from the Fourth Assessment Report in 2007. Why was this particular visualization of the danger of climate change more controversial than other emissions scenarios? How is it different from the other data images in the Third Assessment Report?

PART 5

CLIMATE CHANGE GOVERNANCE

In part, the scientific controversies that arose in the 1980s gained new importance in the 1990s because scientists, environmentalists, and some politicians began to marry climate change discourse to international programs of environmental governance in ways that promised to have far-reaching political and economic impacts. Where liminal scientific literature introduced climate change to broader public and political audiences in the 1970s and 1980s, in the 1990s scientists worked with politicians and foreign governments to create liminal scientific *institutions*. Known to political scientists as "boundary organizations," these institutions were meant to serve as links between scientists' understanding of climate change and national and international climate change policy.[1] The documents they produced, like the institutions themselves, straddled the divide between science and politics. A blend of scientific explanation, international legal nicety, and conflicting geopolitical objectives, these documents both reflected and helped shape the dual scientific and political processes behind their creation.

The most important boundary organization in climate change politics since the early 1990s has been the Intergovernmental Panel on Climate Change (IPCC). Founded under the auspices of the United Nations and the World Meteorological Organization in 1988, the IPCC brought together scientists and government representatives from around the world to create a broad consensus on the state of climate change science that could serve as a baseline for negotiations on an international climate change treaty, the United Nations Framework Convention on Climate Change (UNFCCC). Based on the IPCC's First Assessment Report from 1990 (variously known as FAR, AR1, and IPCC 1990), the treaty would then be implemented by a series of protocols (Kyoto the most famous among them) that continued to rely on updated IPCC assessments, produced every five to seven years. The IPCC was never actually intended to conduct science; rather, independently conducted science, funneled through the consensus-building mechanism of the IPCC, would inform international climate change policy. (The IPCC has now released a total of five comprehensive assessment reports).

As a document, however, an assessment like IPCC 1990 or its successor, IPCC 1995, has some idiosyncratic characteristics. Like other institutional documents, an assessment report does not have one author; rather, the final document bears the imprint of hundreds of scientists and government representatives from around the world. As an organization, the IPCC consisted (and still consists) primarily of three main working groups—Working Group 1 on basic science, Working Group 2 on impacts and adaptation, and Working Group 3 on mitigation and policy responses—each supported in the drafting process by a Technical Support Unit. The working groups rely on teams of lead authors, each team in charge of a particular chapter tackling its particular area of expertise. Author teams work together to compile and summarize up-to-date scientific understanding on specific topics based on peer-reviewed scientific literature on those topics. Lead authors then use those summaries to construct chapters that are subsequently reviewed by technical experts, organizations, and governments. Revised drafts are then incorporated into a "summary for policymakers," to be approved by the working group and then revised again for submission to a plenary of government representatives to the IPCC (many scientifically trained, but some less so). The summaries for policymakers (SPMs) are the most important and controversial products of the IPCC. The plenary debates the summaries line by line, each delegate lobbying for specific language that reflects a particular country's political interests. The process can produce slippery but significant terms of art with rich and multifaceted political meanings—like "discernible human influence," which appeared after intense negotiations in point 4 in the WG1 Summary for Policymakers from 1995.

The scientific content of an IPCC assessment represents a sort of snapshot of scientific consensus in a particular moment, and one marker of the organization's success is that each assessment informs the political context of nearly all forms of climate change discourse until the next assessment comes out. IPCC assessments thus serve as valuable contextual touchstones for reading other historical documents like the text of the UNFCCC or, much later, the so-called Stern Review on the economics of climate change, put out by the British government in 2006.

But the IPCC process itself has also been shaped by the evolving politics of climate change. The science-to-politics avenue is actually a two-way street. Reading IPCC assessment reports in conversation both with each other and with documents from the political negotiations they inform provides a way to think about how both the consensus-making process and the mechanisms of climate governance have changed over time. IPCC reports reflect developments in scientific understanding over five- to seven-year intervals, but changes in the IPCC process, the style of its publications, and the focus of its discussions also reflect the particular interests of its intended users: the international political community. In 1995, for example, the Second Assessment Report provided a detailed description of the IPCC process, now including peer review, as a response to criticism that the First Assessment Report lacked transparency and rigor. It also referred directly to the objectives of the UNFCCC, a treaty that did not yet exist in 1990 when the First Assessment Report was released. Six years later, in 2001, the Third Assessment Report built on the Second Assessment Report's engagement with the UNFCCC in a more robust "synthesis report," a sort of grand executive summary that linked the basic physical understanding of the climate system to the societal impacts and potential responses to climate change in order to emphasize the report's "policy relevance" for political actors, only some of whom had signed on to the 1997 Kyoto Protocol.

As it changed, the IPCC and the documents it produced remained embedded in a larger conversation about environmental governance that was also changing in the 1990s and early 2000s. The IPCC self-consciously spoke to the treaty-protocol system of the UNFCCC. As a result, it also implicitly and sometimes explicitly engaged with concerns about economic development, technology transfer, and equity that the UNFCCC inherited from an international push in the late 1980s and early 1990s to tackle environmental problems through international legal and political agreements based on the concept of sustainable development. Defined in 1987 by the United Nations' World Commission on Environment and Development (WCED) as "the kind of development that meets the needs of the present without compromising the ability of future generations to meet their own needs," sustainable development became a key buzzword in international environmental politics in the early 1990s and served as the guiding principle for the United Nations Conference on Environment and Development in Rio de Janeiro, Brazil, where the UNFCCC was introduced. It also served as a bridge between the often conflicting concerns of the developed and developing worlds.

Sustainable development was (and is) a slippery concept, however. Much as government representatives' interpretations of scientific consensus on climate change in IPCC summaries for policymakers reflected their countries' political interests, differing interpretations of sustainable development in practice supported competing objectives in environmental governance. Most importantly, many industrialized nations—particularly the United States—differed from less developed nations in their interpretations of the relative responsibilities of developed and developing nations in funding and implementing sustainable development plans. Borrowing from the Rio Declaration on Environment and Development, the UNFCCC refers to the "common but differentiated" responsibilities on climate change, but nowhere is the tension between the shared responsibilities in mitigating carbon emissions and the differentiated needs of the developing world more clearly evident than in the text of the treaty itself. Not only did the UNFCC pay lip service to developing world needs; the structure of the treaty actually created very different roles for a list of developed ("Annex I") countries, who had clear objectives and responsibilities in emissions reductions, and developing countries, who had few immediate responsibilities under the first iteration of the treaty. (Introduced at a transformative moment in history at the end of the Cold War, the UNFCC also recognized that some "industrialized" countries in the former Soviet Union—called "economies in transition"—did not have the resources to contribute to mitigation in the near term and ought to be treated separately.) Five years later, this same concern over the role of the developing world in mitigating climate change would serve as a locus of dissent in the development of the Kyoto Protocol—a protocol significantly influenced by U.S. interests, shaped by U.S. negotiators (including Vice President Al Gore), and signed by President Bill Clinton, but a protocol that was never introduced in the U.S. Senate, which preemptively resolved to reject it in the unanimous 1997 Byrd-Hagel Resolution.

With so many overlapping and dynamic elements involved in climate change governance, how are we as historians to make sense of the documents created through the complex processes of scientific and governmental organizations? How do we read climate change in context in the 1990s?

One way to approach documents created by organizations involved in governance is to think about them metaphorically as something like a set of Russian nesting dolls. Like any individual doll in a Matryoshka set, a document like an IPCC assessment report has a variety of unique features characteristic of its unique objectives within the processes of environmental governance. An IPCC report can, like a single doll, be evaluated independently as a snapshot of scientific consensus, and it can be compared with other assessments to see how scientific consensus and the process of creating it has changed over time. How would you characterize the structure and tone of the First Assessment Report? How does it differ from the Third Assessment Report, produced in 2001?

As with a single doll, however, an individual organizational document's form also reflects its position within the larger set. Each organization produces documents that engage with an internal politics—the smaller dolls inside—and an external politics, the larger dolls outside. Though identical to neither a peer-reviewed scientific article nor a legal treaty, the IPCC necessarily conforms to both the scientific literature it comprises and the legal process it supports. What elements of the First Assessment Report do you see in the UNFCCC that help translate science into a meaningful form for the international political arena? How does the Second Assessment Report (IPCC 1995) borrow from or satisfy both the scientific and political discourses it represents?

Taking documents from various parts of the governance process together—from the WCED report down through the scientific literature supporting chapters within IPCC working groups—can provide a sense of the common assumptions and characteristic disagreements that have shaped the process as a whole as it has changed over time. Where do documents from different parts of the sustainability-focused climate change governance process overlap, and what does that say about the assumptions, priorities, and liabilities of climate change governance in the 1990s?

NOTE

1   Stephen Bocking, *Nature's Experts: Science, Politics, and the Environment* (New Brunswick, NJ: Rutgers University Press, 2004); David H. Guston, "Boundary Organizations in Environmental Policy and Science: An Introduction," *Science, Technology, and Human Values* 26, no. 4 (October 2001): 399–408.

INTERGOVERNMENTAL PANEL ON CLIMATE CHANGE

FIRST ASSESSMENT REPORT

Released in 1990, the First Assessment Report (AR1) of the Intergovernmental Panel on Climate Change (IPCC) presented a relatively straightforward picture of what scientists did and did not know about the global climate system at the time. To its participants, the political implications of the IPCC had already become clear by the time AR1 was released, but in 1990 the process included significantly fewer scientists and administrators than later iterations would, and the executive summary of Working Group 1—at that point by far the most important and prestigious working group—reflected the scientific objectives that guided the study. In the context of other statements on climate change, AR1 was hardly a call to arms; in fact, the IPCC expressly contested the assertion by NASA climate scientist James Hansen in 1988 that models could already detect the impacts of climate change, taking a more conservative approach to detecting climate impacts. Even so, the report reaffirmed with confidence the probability of future greenhouse warming, providing a strong scientific basis upon which to build a political agreement on climate change.

. . . .

J. T. Houghton, G. J. Jenkins, and J. J. Ephraums, eds., *Climate Change: The IPCC Scientific Assessment (1990)* (Cambridge: Cambridge University Press, 1990). Excerpt: xi–xii.

EXECUTIVE SUMMARY

We are certain of the following:

- there is a natural greenhouse effect which already keeps the Earth warmer than it would otherwise be.
- emissions resulting from human activities are substantially increasing the atmospheric concentrations of the greenhouse gases carbon dioxide, methane, chlorofluorocarbons (CFCs) and nitrous oxide. These increases will enhance the greenhouse effect, resulting on average in an additional warming of the Earth's surface. The main greenhouse gas, water vapour, will increase in response to global warming and further enhance it.

We calculate with confidence that:

- some gases are potentially more effective than others at changing climate, and their relative effectiveness can be estimated. Carbon dioxide has been responsible for over half the enhanced greenhouse effect in the past, and is likely to remain so in the future.
- atmospheric concentrations of the long-lived gases (carbon dioxide, nitrous oxide and the CFCs) adjust only slowly to changes in emissions. Continued emissions of these gases at present rates would commit us to increased concentrations for centuries ahead. The longer emissions continue to increase at present day rates, the greater reductions would have to be for concentrations to stabilise at a given level.
- the long-lived gases would require immediate reductions in emissions from human activities of over 60% to stabilise their concentrations at today's levels, methane would require a 15–20% reduction.

Based on current model results, we predict:

- under the IPCC Business-as-Usual (Scenario A) emissions of greenhouse gases, a rate of increase of global mean temperature during the next century of about 0.3°C per decade (with an uncertainty range of 0.2°C to 0.5°C per decade), this is greater than that seen over the past 10,000 years. This will result in a likely increase

in global mean temperature of about 1°C above the present value by 2025 and 3°C before the end of the next century. The rise will not be steady because of the influence of other factors.
- under the other IPCC emission scenarios, which assume progressively increasing levels of controls, rates of increase in global mean temperature of about 0.2°C per decade (Scenario B), just above 0.1°C per decade (Scenario C) and about 0.1°C per decade (Scenario D).
- that land surfaces warm more rapidly than the ocean, and high northern latitudes warm more than the global mean in winter.
- regional climate changes different from the global mean, although our confidence in the prediction of the detail of regional changes is low. For example, temperature increases in Southern Europe and central North America are predicted to be higher than the global mean, accompanied on average by reduced summer precipitation and soil moisture. There are less consistent predictions for the tropics and the Southern Hemisphere.
- under the IPCC Business-as-Usual emissions scenario, an average rate of global mean sea level rise of about 6cm per decade over the next century (with an uncertainty range of 3–10cm per decade) mainly due to thermal expansion of the oceans and the melting of some land ice. The predicted rise is about 20cm in global mean sea level by 2030, and 65cm by the end of the next century. There will be significant regional variations.

There are many uncertainties in our predictions, particularly with regard to the timing, magnitude and regional patterns of climate change, due to our incomplete understanding of:

- sources and sinks of greenhouse gases, which affect predictions of future concentrations
- clouds, which strongly influence the magnitude of climate change
- oceans, which influence the timing and patterns of climate change
- polar ice sheets, which affect predictions of sea level rise

These processes are already partially understood, and we are confident that the uncertainties can be reduced by further research. However, the complexity of the system means that we cannot rule out surprises.

Our judgment is that:

- Global mean surface air temperature has increased by 0.3°C to 0.6°C over the last 100 years, with the five global-average warmest years being in the 1980s. Over the same period global sea level has increased by 10–20cm. These increases have not been smooth with time, nor uniform over the globe.
- The size of this warming is broadly consistent with predictions of climate models, but it is also of the same magnitude as natural climate variability. Thus the observed increase could be largely due to this natural variability, alternatively this variability and other human factors could have offset a still larger human-induced greenhouse warming. The unequivocal detection of the enhanced greenhouse effect from observations is not likely for a decade or more.
- There is no firm evidence that climate has become more variable over the last few decades. However, with an increase in the mean temperature, episodes of high temperatures will most likely become more frequent in the future, and cold episodes less frequent.
- Ecosystems affect climate, and will be affected by a changing climate and by increasing carbon dioxide concentrations. Rapid changes in climate will change the composition of ecosystems, some species will benefit while others will be unable to migrate or adapt fast enough and may become extinct. Enhanced levels of carbon dioxide may increase productivity and efficiency of water use of vegetation. The effect of warming on biological processes, although poorly understood, may increase the atmospheric concentrations of natural greenhouse gases.

To improve our predictive capability, we need:

- to understand better the various climate-related processes, particularly those associated with clouds, oceans and the carbon cycle
- to improve the systematic observation of climate-related variables on a global basis, and further investigate changes which took place in the past

- to develop improved models of the Earth's climate system
- to increase support for national and international climate research activities, especially in developing countries
- to facilitate international exchange of climate data

WORLD COMMISSION ON
ENVIRONMENT AND DEVELOPMENT

OUR COMMON FUTURE
(THE BRUNDTLAND REPORT)

The IPCC undertook its First Assessment Report amid the construction of a new paradigm in international environmental governance: that of "sustainable development." Codified by the World Commission on Environment and Development in the late 1980s as "development that meets the needs of the present without compromising the ability of future generations to meet their own needs," the concept of sustainable development reflected the context of its creation. Between the end of the Second World War and the beginning of the 1980s, organs of international finance like the World Bank and the International Monetary Fund ushered in a period of financial liberalization that saw a new globalization of capital as investors in the developed world pumped money into resource extraction in the less developed world. By the mid-1980s, however, the social and environmental consequences of unregulated private investment had become increasingly clear and dire for the people of developing nations. Sensitive to the charges of exploitation from less developed member states, in 1983 the United Nations General Assembly commissioned Norwegian prime minister Gro Harlem Brundtland and a group of UN experts to articulate a vision for economic growth in the developing world that at once rejected Soviet-style central planning and addressed the excesses of global capitalism. The result, a report called

---

World Commission on Environment and Development, *Our Common Future: Report of the World Commission on Environment and Development* (Oxford: Oxford University Press, 1987). Excerpt: pp. 1, 11, 12–13, 18, 54. Also available as United Nations document A/42/427, www.un-documents.net/wced-ocf.htm. Copyright 1987, United Nations. Reprinted with the permission of the United Nations.

Our Common Future, published in 1987 and colloquially known as the Brundtland Report, articulated a concept of sustainable development that would guide international environmental politics—and especially the politics of climate change—for the next three decades.*

· ·

NOTE BY THE SECRETARY-GENERAL

1. The General Assembly, in its resolution 38/161 of 19 December 1983, *inter alia*, welcomed the establishment of a special commission that should make available a report on environment and the global *problematique* to the year 2000 and beyond, including proposed strategies for sustainable development. The commission later adopted the name World Commission on Environment and Development. In the same resolution, the Assembly decided that, on matters within the mandate and purview of the United Nations Environment Programme, the report of the special commission should in the first instance be considered by the Governing Council of the Programme, for transmission to the Assembly together with its comments, and for use as basic material in the preparation, for adoption by the Assembly, of the Environmental Perspective to the Year 2000 and Beyond.

CHAIRMAN'S FOREWORD

"A global agenda for change"—this was what the World Commission on Environment and Development was asked to formulate. It was an urgent call by the General Assembly of the United Nations:

- to propose long-term environmental strategies for achieving sustainable development by the year 2000 and beyond;
- to recommend ways concern for the environment may be translated into greater co-operation among developing countries and

---

* For a readable account of the IMF and World Bank in international governance, see "Development as World-Making" and "The Real New International Economic Order," in Mark Mazower, *Governing the World: The History of an Idea, 1815 to the Present* (New York: Penguin, 2012), 273–304, 343–77.

between countries at different stages of economical and social development and lead to the achievement of common and mutually supportive objectives that take account of the interrelationships between people, resources, environment, and development;
- to consider ways and means by which the international community can deal more effectively with environment concerns; and
- to help define shared perceptions of long-term environmental issues and the appropriate efforts needed to deal successfully with the problems of protecting and enhancing the environment, a long-term agenda for action during the coming decades, and aspirational goals for the world community. . . .

Perhaps our most urgent task today is to persuade nations of the need to return to multilateralism. The challenge of reconstruction after the Second World War was the real motivating power behind the establishment of our post-war international economic system. The challenge of finding sustainable development paths ought to provide the impetus—indeed the imperative—for a renewed search for multilateral solutions and a restructured international economic system of co-operation. These challenges cut across the divides of national sovereignty, of limited strategies for economic gain, and of separated disciplines of science.

After a decade and a half of a standstill or even deterioration in global cooperation, I believe the time has come for higher expectations, for common goals pursued together, for an increased political will to address our common future. . . .

The present decade has been marked by a retreat from social concerns. Scientists bring to our attention urgent but complex problems bearing on our very survival: a warming globe, threats to the Earth's ozone layer, deserts consuming agricultural land. We respond by demanding more details, and by assigning the problems to institutions ill equipped to cope with them. Environmental degradation, first seen as mainly a problem of the rich nations and a side effect of industrial wealth, has become a survival issue for developing nations. It is part of the downward spiral of linked ecological and economic decline in which many of the poorest nations are trapped. Despite official hope expressed on all sides, no trends identifiable today, no programmes or policies, offer any real hope of narrowing the growing gap between rich and poor nations. And as part of our "development," we have amassed weapons arsenals capable of diverting the paths that evolution has followed for millions of years and of creating a planet our ancestors would not recognize....

FROM ONE EARTH TO ONE WORLD

... This Commission believes that people can build a future that is more prosperous, more just, and more secure. Our report, *Our Common Future*, is not a prediction of ever increasing environmental decay, poverty, and hardship in an ever more polluted world among ever decreasing resources. We see instead the possibility for a new era of economic growth, one that must be based on policies that sustain and expand the environmental resource base. And we believe such growth to be absolutely essential to relieve the great poverty that is deepening in much of the developing world....

TOWARDS SUSTAINABLE DEVELOPMENT

1. Sustainable development is development that meets the needs of the present without compromising the ability of future generations to meet their own needs. It contains within it two key concepts:

- the concept of 'needs,' in particular the essential needs of the world's poor, to which overriding priority should be given; and
- the idea of limitations imposed by the state of technology and social organization on the environment's ability to meet present and future needs.

2. Thus the goals of economic and social development must be defined in terms of sustainability in all countries developed or developing, market-oriented or centrally planned. Interpretations will vary, but must share certain general features and must flow from a consensus on the basic concept of sustainable development and on a broad strategic framework for achieving it.

UNITED NATIONS

RIO DECLARATION ON ENVIRONMENT AND DEVELOPMENT

The publication of the Brundtland Report added momentum to a large-scale effort to define and codify the principles of sustainable development already underway within the international community, and no event drew more focus in this discussion than the 1992 United Nations Conference on Environment and Development (UNCED) in Rio de Janeiro, Brazil. Commonly called the Earth Summit, Rio saw the introduction of a variety of treaties and proposals meant to curb environmental degradation and promote an inclusive, internationalist vision of sustainable development, including the United Nations Framework Convention on Climate Change. The guiding document and key ideological product of the conference, however, was the Rio Declaration on Environment and Development, which laid out the principles of sustainable development as negotiated and adopted by the international political community. What similarities between the documents tie the Brundtland Report to the Rio Declaration? How do the documents differ? What changed between 1987 and 1992 that might complicate the continuity of the concept of sustainable development from the Brundtland Report to the UNCED document?*

United Nations, "Rio Declaration on Environment and Development" (1992). Excerpt: pg. 1–2, 3, 4–5. Available at www.unesco.org/education/nfsunesco/pdf/RIO_E.PDF. Copyright 1992 United Nations. Reprinted with the permission of the United Nations.

* For an excellent collection of documents from the Earth Summit, see Ken Conca et al., *Green Planet Blues: Environmental Politics from Stockholm to Rio* (Boulder, CO: Westview Press, 1995).

PREAMBLE

The United Nations Conference on Environment and Development, Having met at Rio de Janeiro from 3 to 14 June 1992,

Reaffirming the Declaration of the United Nations Conference on the Human Environment, adopted at Stockholm on 16 June 1972, and seeking to build upon it,

With the goal of establishing a new and equitable global partnership through the creation of new levels of co-operation among States, key sectors of societies and people,

Working towards international agreements which respect the interests of all and protect the integrity of the global environmental and developmental system,

Recognizing the integral and interdependent nature of the Earth, our home,

Proclaims that:

PRINCIPLE 1

Human beings are at the centre of concerns for sustainable development. They are entitled to a healthy and productive life in harmony with nature.

PRINCIPLE 2

States have, in accordance with the Charter of the United Nations and the principles of international law, the sovereign right to exploit their own resources pursuant to their own environmental and developmental policies, and the responsibility to ensure that activities within their jurisdiction or control do not cause damage to the environment of other States or of areas beyond the limits of national jurisdiction.

PRINCIPLE 3

The right to development must be fulfilled so as to equitably meet developmental and environmental needs of present and future generations.

PRINCIPLE 4

In order to achieve sustainable development, environmental protection shall constitute an integral part of the development process and cannot be considered in isolation from it.

PRINCIPLE 5

All States and all people shall cooperate in the essential task of eradicating poverty as an indispensable requirement for sustainable development, in order to decrease the disparities in standards of living and better meet the needs of the majority of the people of the world.

PRINCIPLE 6

The special situation and needs of developing countries, particularly the least developed and those most environmentally vulnerable, shall be given special priority. International actions in the field of environment and development should also address the interests and needs of all countries.

PRINCIPLE 7

States shall cooperate in a spirit of global partnership to conserve, protect and restore the health and integrity of the Earth's ecosystem. In view of the different contributions to global environmental degradation, States have common but differentiated responsibilities.

The developed countries acknowledge the responsibility that they bear in the international pursuit of sustainable development in view of the pressures their societies place on the global environment and of the technologies and financial resources they command. . . .

PRINCIPLE 12

States should cooperate to promote a supportive and open international economic system that would lead to economic growth and sustainable development in all countries, to better address the problems of environmental degradation. Trade policy measures for environmental purposes should not constitute a means of arbitrary or unjustifiable discrimination or a disguised restriction on international trade. Unilateral actions to deal with environmental challenges outside the jurisdiction of the importing country should be avoided. Environmental measures addressing transboundary or global environmental problems should, as far as possible, be based on an international consensus. . . .

PRINCIPLE 16

National authorities should endeavour to promote the internalization of environmental costs and the use of economic instruments, taking into account the approach that the polluter should, in principle, bear the cost of pollution, with due regard to the public interest and without distorting international trade and investment. . . .

PRINCIPLE 20

Women have a vital role in environmental management and development. Their full participation is therefore essential to achieve sustainable development.

PRINCIPLE 21

The creativity, ideals and courage of the youth of the world should be mobilized to forge a global partnership in order to achieve sustainable development and ensure a better future for all.

PRINCIPLE 22

Indigenous people and their communities, and other local communities, have a vital role in environmental management and development because of their knowledge and traditional practices. States should recognize and duly support their identity, culture and interests and enable their effective participation in the achievement of sustainable development.

PRINCIPLE 23

The environment and natural resources of people under oppression, domination and occupation shall be protected.

PRINCIPLE 24

Warfare is inherently destructive of sustainable development. States shall therefore respect international law providing protection for the environment in times of armed conflict and cooperate in its further development, as necessary.

PRINCIPLE 25

Peace, development and environmental protection are interdependent and indivisible.

UNITED NATIONS

UNITED NATIONS FRAMEWORK CONVENTION ON CLIMATE CHANGE (UNFCCC)

Introduced in Rio in 1992, the United Nations Framework Convention on Climate Change (UNFCCC, or "U-N-F-Triple-C") was the first international legal treaty written specifically to deal with the problem of climate change, and it has stood as the primary framework for international climate change governance since its ratification in 1993. The UNFCCC follows the conventions of an international legal treaty. It comprises a preamble; a list of definitions; a stated objective; twenty-six articles on the rules, principles, and procedures of the agreement; and two annexes delineating different nations responsible for different actions under the stipulations of the treaty. Its content also reflects the context in which it was created. Where is there consonance between the UNFCCC and the Rio Declaration? What is different between the two documents? How does the abiding concern over the "common but differentiated responsibilities" of the parties involved in the UNFCCC align with the Rio Declaration's commitment to the interests of the developing world in environmental governance? Though very different in function, both were born of the same post–Cold War Earth Summit milieu. What common themes draw them together?

• •

United Nations Framework Convention on Climate Change (May 9, 1992), S. Treaty Doc No. 102-38, 1771 U.N.T.S. 107. Excerpt: pp. 1–3, 4–6, 8, 20, 23–24. Copyright 1992 United Nations. Reprinted with the permission of the United Nations.

The Parties to this Convention,

Acknowledging that change in the Earth's climate and its adverse effects are a common concern of humankind,

Concerned that human activities have been substantially increasing the atmospheric concentrations of greenhouse gases, that these increases enhance the natural greenhouse effect, and that this will result on average in an additional warming of the Earth's surface and atmosphere and may adversely affect natural ecosystems and humankind,

Noting that the largest share of historical and current global emissions of greenhouse gases has originated in developed countries, that per capita emissions in developing countries are still relatively low and that the share of global emissions originating in developing countries will grow to meet their social and development needs,

Aware of the role and importance in terrestrial and marine ecosystems of sinks and reservoirs of greenhouse gases,

Noting that there are many uncertainties in predictions of climate change, particularly with regard to the timing, magnitude and regional patterns thereof,

Acknowledging that the global nature of climate change calls for the widest possible cooperation by all countries and their participation in an effective and appropriate international response, in accordance with their common but differentiated responsibilities and respective capabilities and their social and economic conditions,

Recalling the pertinent provisions of the Declaration of the United Nations Conference on the Human Environment, adopted at Stockholm on 16 June 1972,

Recalling also that States have, in accordance with the Charter of the United Nations and the principles of international law, the sovereign right to exploit their own resources pursuant to their own environmental and developmental policies, and the responsibility to ensure that activities within their jurisdiction or control do not cause damage to the environment of other States or of areas beyond the limits of national jurisdiction,

Reaffirming the principle of sovereignty of States in international cooperation to address climate change,

Recognizing that States should enact effective environmental legislation, that environmental standards, management objectives and priorities should reflect the environmental and developmental context to which they apply, and that standards applied by some countries may be inappropriate and of unwarranted economic and social cost to other countries, in particular developing countries,

Recalling the provisions of General Assembly resolution 44/228 of 22 December 1989 on the United Nations Conference on Environment and Development, and resolutions 43/53 of 6 December 1988, 44/207 of 22 December 1989, 45/212 of 21 December 1990 and 46/169 of 19 December 1991 on protection of global climate for present and future generations of mankind,

Recalling also the provisions of General Assembly resolution 44/206 of 22 December 1989 on the possible adverse effects of sea-level rise on islands and coastal areas, particularly low-lying coastal areas and the pertinent provisions of General Assembly resolution 44/172 of 19 December 1989 on the implementation of the Plan of Action to Combat Desertification,

Recalling further the Vienna Convention for the Protection of the Ozone Layer, 1985, and the Montreal Protocol on Substances that Deplete the Ozone Layer, 1987, as adjusted and amended on 29 June 1990,

Noting the Ministerial Declaration of the Second World Climate Conference adopted on 7 November 1990,

Conscious of the valuable analytical work being conducted by many States on climate change and of the important contributions of the World Meteorological Organization, the United Nations Environment Programme and other organs, organizations and bodies of the United Nations system, as well as other international and intergovernmental bodies, to the exchange of results of scientific research and the coordination of research,

Recognizing that steps required to understand and address climate change will be environmentally, socially and economically most effective if they are based on relevant scientific, technical and economic considerations and continually re-evaluated in the light of new findings in these areas,

Recognizing that various actions to address climate change can be justified economically in their own right and can also help in solving other environmental problems,

Recognizing also the need for developed countries to take immediate action in a flexible manner on the basis of clear priorities, as a first step towards comprehensive response strategies at the global, national and, where agreed, regional levels that take into account all greenhouse gases, with due consideration of their relative contributions to the enhancement of the greenhouse effect,

Recognizing further that low-lying and other small island countries, countries with low-lying coastal, arid and semi-arid areas or areas liable to floods, drought and desertification, and developing countries with fragile mountainous ecosystems are particularly vulnerable to the adverse effects of climate change,

Recognizing the special difficulties of those countries, especially developing countries, whose economies are particularly dependent on fossil fuel production, use and exportation, as a consequence of action taken on limiting greenhouse gas emissions,

Affirming that responses to climate change should be coordinated with social and economic development in an integrated manner with a view to avoiding adverse impacts on the latter, taking into full account the legitimate priority needs of developing countries for the achievement of sustained economic growth and the eradication of poverty,

Recognizing that all countries, especially developing countries, need access to resources required to achieve sustainable social and economic development and that, in order for developing countries to progress towards that goal, their energy consumption will need to grow taking into account the possibilities for achieving greater energy efficiency and for controlling greenhouse gas emissions in general, including through the application of new technologies on terms which make such an application economically and socially beneficial,

Determined to protect the climate system for present and future generations,

Have agreed as follows: . . .

ARTICLE 2: OBJECTIVE

The ultimate objective of this Convention and any related legal instruments that the Conference of the Parties may adopt is to achieve, in accordance with the relevant provisions of the Convention, stabilization of greenhouse gas concentrations in the atmosphere at a level that would prevent dangerous anthropogenic interference with the climate system. Such a level should be achieved within a time frame sufficient to allow ecosystems to adapt naturally to climate change, to ensure that food production is not threatened and to enable economic development to proceed in a sustainable manner.

ARTICLE 3: PRINCIPLES

In their actions to achieve the objective of the Convention and to implement its provisions, the Parties shall be guided, inter alia, by the following:

1. The Parties should protect the climate system for the benefit of present and future generations of humankind, on the basis of equity and in accordance with their common but differentiated responsibilities and respective capabilities. Accordingly, the developed country Parties should take the lead in combating climate change and the adverse effects thereof.
2. The specific needs and special circumstances of developing country Parties, especially those that are particularly vulnerable to the adverse effects of climate change, and of those Parties, especially developing country Parties, that would have to bear a disproportionate or abnormal burden under the Convention, should be given full consideration.
3. The Parties should take precautionary measures to anticipate, prevent or minimize the causes of climate change and mitigate its adverse effects. Where there are threats of serious or irreversible damage, lack of full scientific certainty should not be used as a reason for postponing such measures, taking into account that policies and measures to deal with climate change should be cost-effective so as to ensure global benefits at the lowest possible cost. . . .
4. The Parties have a right to, and should, promote sustainable development. . . .

ARTICLE 4: COMMITMENTS

1. All Parties, taking into account their common but differentiated responsibilities and their specific national and regional development priorities, objectives and circumstances, shall:

    (a) Develop, periodically update, publish and make available to the Conference of the Parties, in accordance with Article 12, national inventories of anthropogenic emissions by sources and removals by sinks of all greenhouse gases not controlled by the Montreal Protocol, using comparable methodologies to be agreed upon by the Conference of the Parties;

    (b) Formulate, implement, publish and regularly update national and, where appropriate, regional programmes containing measures to mitigate climate change by addressing anthropogenic emissions by sources and removals by sinks of all greenhouse gases not controlled by the Montreal Protocol, and measures to facilitate adequate adaptation to climate change;

    (c) Promote and cooperate in the development, application and diffusion, including transfer, of technologies, practices and processes that control, reduce or prevent anthropogenic emissions of greenhouse gases not controlled by the Montreal Protocol in all relevant sectors, including the energy, transport, industry, agriculture, forestry and waste management sectors;

    (d) Promote sustainable management, and promote and cooperate in the conservation and enhancement, as appropriate, of sinks and reservoirs of all greenhouse gases not controlled by the Montreal Protocol, including biomass, forests and oceans as well as other terrestrial, coastal and marine ecosystems;

    (e) Cooperate in preparing for adaptation to the impacts of climate change; develop and elaborate appropriate and integrated plans for coastal zone management, water resources and agriculture, and for the protection and rehabilitation of areas, particularly in Africa, affected by drought and desertification, as well as floods;

    (f) Take climate change considerations into account, to the extent feasible, in their relevant social, economic and environmental policies and actions, and employ appropriate methods, for example impact assessments, formulated and determined nationally, with a view to minimizing adverse effects on the economy, on public health and on the quality of the environment, of projects or measures undertaken by them to mitigate or adapt to climate change;

(g) Promote and cooperate in scientific, technological, technical, socio-economic and other research, systematic observation and development of data archives related to the climate system and intended to further the understanding and to reduce or eliminate the remaining uncertainties regarding the causes, effects, magnitude and timing of climate change and the economic and social consequences of various response strategies; ...

2. The developed country Parties and other Parties included in Annex I commit themselves specifically as provided for in the following:

(a) Each of these Parties shall adopt national policies and take corresponding measures on the mitigation of climate change, by limiting its anthropogenic emissions of greenhouse gases and protecting and enhancing its greenhouse gas sinks and reservoirs. . . . These Parties may implement such policies and measures jointly with other Parties and may assist other Parties in contributing to the achievement of the objective of the Convention and, in particular, that of this subparagraph; ...

3. The developed country Parties and other developed Parties included in Annex II shall provide new and additional financial resources to meet the agreed full costs incurred by developing country Parties in complying with their obligations under Article 12, paragraph 1. They shall also provide such financial resources, including for the transfer of technology, needed by the developing country Parties to meet the agreed full incremental costs of implementing measures that are covered by paragraph 1 of this Article and that are agreed between a developing country Party and the international entity or entities referred to in Article 11, in accordance with that Article. The implementation of these commitments shall take into account the need for adequacy and predictability in the flow of funds and the importance of appropriate burden sharing among the developed country Parties.

4. The developed country Parties and other developed Parties included in Annex II shall also assist the developing country Parties that are

particularly vulnerable to the adverse effects of climate change in meeting costs of adaptation to those adverse effects.

5. The developed country Parties and other developed Parties included in Annex II shall take all practicable steps to promote, facilitate and finance, as appropriate, the transfer of, or access to, environmentally sound technologies and know-how to other Parties, particularly developing country Parties, to enable them to implement the provisions of the Convention. . . .

6. In the implementation of their commitments under paragraph 2 above, a certain degree of flexibility shall be allowed by the Conference of the Parties to the Parties included in Annex I undergoing the process of transition to a market economy, in order to enhance the ability of these Parties to address climate change, including with regard to the historical level of anthropogenic emissions of greenhouse gases not controlled by the Montreal Protocol chosen as a reference.

7. The extent to which developing country Parties will effectively implement their commitments under the Convention will depend on the effective implementation by developed country Parties of their commitments under the Convention related to financial resources and transfer of technology and will take fully into account that economic and social development and poverty eradication are the first and overriding priorities of the developing country Parties. . . .

ARTICLE 20: SIGNATURE

This Convention shall be open for signature by States Members of the United Nations or of any of its specialized agencies or that are Parties to the Statute of the International Court of Justice and by regional economic integration organizations at Rio de Janeiro, during the United Nations Conference on Environment and Development, and thereafter at United Nations Headquarters in New York from 20 June 1992 to 19 June 1993. . . .

ANNEX

Australia
Austria
Belarus
Belgium
Bulgaria
Canada
Croatia
Czech Republic
Denmark
European Economic Community
Estonia
Finland
France
Germany
Greece
Hungary
Iceland
Ireland
Italy
Japan
Latvia
Liechtenstein
Lithuania
Luxembourg
Monaco
Netherlands
New Zealand
Norway
Poland
Portugal
Romania
Russian Federation
Slovakia
Slovenia
Spain
Sweden
Switzerland
Turkey
Ukraine
United Kingdom of Great Britain and Northern Ireland
United States of America

ANNEX II*

Australia
Austria
Belgium
Canada
Denmark
European Economic Community
Finland
France
Germany
Greece
Iceland
Ireland
Italy
Japan
Luxembourg
Netherlands
New Zealand
Norway
Portugal
Spain
Sweden
Switzerland
United Kingdom of Great Britain and Northern Ireland
United States of America

* Turkey was deleted from Annex II by an amendment that entered into force 28 June 2002, pursuant to decision 26/CP.7 adopted at COP 7.

C. BOYDEN GRAY AND DAVID B. RIVKIN JR.

A 'NO REGRETS' ENVIRONMENTAL POLICY

In 1988, then Vice President George H. W. Bush had embraced a moderate, conservation-minded approach to environmental issues. As a presidential candidate in the 1988 election, Bush promised to serve as an "environmental president" and vowed to combat the greenhouse effect with "the White House effect." Once in office, Bush helped pass an evenhanded and remarkably effective new Clean Air Act in 1990, and he sought to build off this success in international environmental governance as negotiations over the United Nations Framework Convention on Climate Change began to heat up. At the same time, however, the administration hoped to avoid radical regulatory steps that might curb American economic growth or limit American trade flexibility. In retrospect, Bush's positions on regulation and environmental protection seem at odds with each other, in part because contemporary debates about climate change and environmental governance have polarized these issues in new and pervasive ways. At the time, however, Bush's conservation strategy made economic and political sense, and it resonated with many moderates, Democrats and Republicans alike. Even so, Bush's particular form of environmentalism required some strategic political thinking. Here, White House counsel Boyden Gray and the Department of Energy's David Rivkin Jr. articulate what became known as Bush's "no regrets" approach to the environment, a creative—and, some might say, cynical—take on sustainable development. In addition to advocating for a broader approach to environmental issues that would focus on whole ecosystems rather than

---

C. Boyden Gray and David B. Rivkin Jr., "A 'No Regrets' Environmental Policy," *Foreign Policy*, no. 83 (July 1991). Excerpt: pp. 51, 52, 62–63.

carbon emissions per se, Bush's aides sought to reframe global warming in terms of the many noncarbon greenhouse gases that the United States had already committed to regulating under the Clean Air Act. By accounting for these gases, alongside the carbon-fixing potential of United States' vast forests, Gray and Rivkin painted a picture of U.S. emissions that actually required very little of American producers or consumers to meet the proposed targets for the UNFCCC.

• •

BUSH ADMINISTRATION POLICY

The Bush administration, despite the claims of global warming doomsayers, is not ignoring this issue. Uncertain science and divisive politics have made for a difficult dilemma: What approach ought to be taken to reduce the allegedly excessive greenhouse effect? There are several possibilities, mostly costly, including adaptation, reforestation, and shifting away from the use of fossil fuels to alternative sources of power. Alternative energy sources offer great promise, and have been aggressively pursued by this administration in the 1990 Clean Air Act Amendments and the National Energy Strategy (NES) legislation. The NES specifically seeks to increase the use of renewable and nonfossil sources of energy. The real issue, however, is the pace of shifts in U.S. energy policy and the extent to which policy should be shaped by global climate change imperatives as opposed to other environmental and energy security considerations. . . .

In view of the uncertainties underlying the global warming debate and the limits on available resources, the administration has embraced a balanced policy of adopting those environmental measures that reduce greenhouse gas emissions while providing concrete environmental benefits. This approach has been termed a "multiple objective steps" or "no regrets" policy. Actions taken in this area should be based upon the long-term outlook, "taking into account the full range of social, economic, and environmental consequences for this and future generations."

Further, a realistic approach to the problem must be a comprehensive one that, based on the best available scientific model of global climate change, reduces all greenhouse gases to the extent necessary by manipulating their differing sources and sinks (carbon-absorbing and -storing elements such as trees or oceans). All greenhouse gases should be addressed, including carbon dioxide, methane, nitrous oxide, halocarbons (which include CFCs and related hydrochlorofluorocarbons) and tropospheric ozone. The complex interplay of sources should be considered as they affect the "net emissions" of greenhouse gases, that in turn influence climate processes. "No regrets" is not tantamount to a "wait and see" approach. As a result of efforts already in place, in the year 2000 the United States will have an aggregate level of greenhouse gas emissions equal to or below the 1987 level. Furthermore, with the enactment of the administration's NES proposals, aggregate levels of greenhouse emissions are projected to remain below the 1987 level through 2030, even accounting for new economic growth. Absolute scientific certainty then is not a prerequisite for taking action. Rather, research should be aggressively pursued and the action taken should make sense based upon the current state of knowledge in this area. . . .

The most important lesson to be drawn from domestic and international experience is that effectively resolving transnational environmental issues requires a comprehensive approach addressing all aspects of pollution at the same time, whether they be air-, water-, or ground-based. . . .

Moreover, any responsible international environmental program must take into account not only strict environmental needs, but also the need for economic and social stability. As EPA Administrator William Reilly pointed out in the Fall 1990 issue of *Policy Review*, without a stable and growing domestic and world economy, the international community will have neither the means, the will, nor the opportunity to overcome current environmental challenges. Thus, sustainable development must advance both economic and environmental values. If the "solution" to global environmental problems works to undermine global political and economic stability, the program may leave the world worse off.

AL GORE AND MITCH McCONNELL

TESTIMONY BEFORE THE SENATE COMMITTEE ON FOREIGN RELATIONS, SEPTEMBER 18, 1992

For the Bush administration, the Rio Earth Summit was a political minefield. In the run-up to the conference, controversies between the developed and developing world—and between the United States and Europe—over the meaning of sustainable development jeopardized agreements on the Rio Declaration, Agenda 21 (a list of actionable sustainability items focused largely on sustainable development in the less developed world), and the UNFCCC. Both the nations of the developing world and Bush's Iraq War allies from the industrialized world looked to the United States to make binding commitments the administration was not prepared to make, and critics on the left promised to make the president's cautious approach to environmental governance an election issue later that year.* When President Bush got to Rio, things got worse. The Brazilian and international press vilified the president for protecting U.S. economic interests over global environmental health, and a memo leaked by a member of Bush's staff undercut the credibility of the administration's chief negotiator, EPA chief William Reilly. The president successfully managed to exclude targets and timetables from the UNFCCC, but he failed to obtain commitments on

---

U.N. Framework Convention on Climate Change: Hearings of Senate Committee on Foreign Relations, 102nd Cong. (1992). Excerpt: pp. 2–3, 17, 18, 19–20, 27–28.

* For more on the run-up to Rio, see Daniel Yergin, "The Road to Rio," in *The Quest: Energy, Security and the Remaking of the Modern World* (New York: Penguin, 2011), 453–70.

emissions reductions from the developing world. Bush signed the treaty on the second-to-last day of the meeting, and almost immediately the Senate began to debate whether to ratify and how to implement the UNFCCC. Here, a testy exchange between Tennessee Democrat (and vice presidential candidate) Al Gore and Kentucky Republican Mitch McConnell on the Senate floor shortly after the Earth Summit reveals some of the nature and terms of the debate. The Senate eventually ratified the UNFCCC in 1993. What role(s) does Al Gore play in this hearing, and how might that affect your reading of his testimony? How does Mitch McConnell play to Gore's multiple roles? Where does this domestic drama fit within the larger international conversation about sustainability that frames the other documents in this chapter?

• •

STATEMENT OF HON. ALBERT GORE, JR., U.S. SENATOR FROM TENNESSEE

Senator GORE. Well thank you so much, Mr. Chairman and Senator Kerry and members of the committee. I am honored to be here and thank you for your generous personal comments. And may I say before I begin my statement to the committee that it was an honor for me to serve as chairman of the Senate's delegation to this Earth Summit. . . .

Mr. Chairman, let me give you the bottomline first and then a brief discussion of why I make the recommendations that I do.

I would urge that we take the following actions: No. 1, we should move quickly and ratify the treaty. No. 2, the United States made a commitment to produce a so-called action agenda on climate change that is called for in the treaty by January 1993. We should initiate immediately a process that is open to the public to produce that document. There should be full opportunity for public comment on the development of that plan, and all background documentation should be made publicly available.

No. 3, we should call for a meeting of the parties to the Convention in January 1993 to review progress. No. 4, since our negotiators have likened the climate treaty to the Vienna Convention, which this committee remembers well, we should make good on that comparison by moving now to develop a protocol to the Convention that will assure meaningful reductions in emissions. We should move forward on this front at the December meeting of the Intergovernmental Negotiating Committee of the Convention.

And No. 5, we need to insure the continued work of the IPCC at its next meeting in November. A subsidiary science and technology body is set up under the Convention. It is imperative that that body merely serve as a liaison to, and in no way displace, the IPCC. We need to have the objective, unpoliticized advice of scientists on these issues, and we can only insure that that will be the case if the independent scientific work of the IPCC is continued.

Now, Mr. Chairman, the process leading to the conclusion of the Climate Change Treaty was initiated and driven by the virtually unanimous opinion of the world scientific community that by overloading the atmosphere with carbon dioxide and other greenhouse gases, we are risking disruptions in the climate balance more severe than any in the past 10,000 years.

Moreover, these disruptions could take place in relatively short time periods over the next several decades. More severe storm systems in some areas, intense bouts of drought in others, rising sea levels and flooding of coastal communities would be some of the consequences predicted by the scientific community. . . .

Apparently all of this, unfortunately, was lost on President Bush. As we all are now all too well aware, the Bush administration was, throughout these negotiations, the single largest obstacle to progress. In the end our intransigence meant that the final agreement is completely devoid of any legally binding commitments to action.

The real meaning of the Earth Summit itself was also lost on President Bush. It was a true turning point in history. Leaders of nearly every nation on Earth gathered together for the first time in a profound awareness of the true nature and magnitude of the global environmental crisis we face.

Perhaps even more significantly, they realized that the alleviation of human suffering around the globe is inextricably intertwined with our efforts to relieve the building pressure on the environment. They understood that to combat the poverty, suffering, and pain that afflicts so many in the world today we have to pursue economic growth that is not harmful to the environment....

I believe the United States of America can create millions of new jobs by leading the environmental revolution, building, manufacturing, selling the new products and technologies that foster economic progress without environmental destruction....

So if we decide to reclaim our heritage, because we have in the past led the world on environmental matters. If we abandon this go slow, protect the polluter approach of the Bush-Quayle administration, we will create a lot of new jobs in the process....

The CHAIRMAN. I thank you very much for being with us and now turn to our ranking Republican.

Senator MCCONNELL. Thank you, Mr. Chairman. Welcome back from the campaign trail, Al. It is good to see you ... we are glad to have you here. After reading a number of your interviews before and after the Earth Summit and reviewing *Earth in the Balance*, it seems to me that you have a very definite vision of the environment. Frankly, I am not sure whether I would call it a nightmare or a hallucination.

I just cannot accept that Americans were embarrassed by the President or that the summit was a disaster waiting to happen. Actually, to put it in your own words as you told the *Post*, you said, "it is like a Greek tragedy, everyone can see it coming. The only thing left is for him"—I guess referring to the President—"is to invite the executives of a few coal companies to come along."

Well, in my State coal means over 100,000 jobs, when you consider direct and coal-related jobs; in the Nation it means 1.1 million jobs more. So they do not welcome that kind of criticism. And, when you told the press and public we cannot wait for all the evidence before we launch the most expensive attack on jobs and economy in this century, frankly, I think the facts are important.

It seems to me that the central criticism you have had of the President, his environmental agenda and this Convention, is the fact that the 156 signatories to the treaty did not agree on a schedule for the reduction of carbon dioxide emissions.

Now, as I understand your position, these carbon dioxide emissions are the root cause of global warming, to you: "the most serious threat we have ever faced." In fact, at least half the scientific community thinks the concern about global warming is greatly exaggerated. At best, many think it is an issue that deserves watching but presents no immediate danger.

But while the problem may be in scientific doubt, you have called for absolute solutions. You demanded at the summit and in legislation, the Global Climate Protection Act, stabilization of carbon dioxide emissions. The question is how do we get there? Clearly the only way to do that, under your proposal would be a tax, a tax on carbon. Never mind the 600,000 workers who would lose their jobs because of this tax. Never mind a 300-percent rise in the price of coal or a 58-percent rise in oil prices. American industries, our farmers, our families, would rather suffer these hard costs than risk the possibility—the possibility of global warming.

I was glad to learn that you had joined a lot of us in agreeing with the President on some of his most important initiatives such as the Clean Air Act. I think it was unanimously viewed as the toughest and most innovative air quality legislation ever passed in our country. But, I must confess I am troubled, and many I know are, by virtually all your other major environmental recommendations. . . .

In any event, we are happy to have you here. [*Laughter*.]. . . .

Senator MCCONNELL. What I am asking you is did you or did you not advocate a carbon tax.

Senator GORE . . . What you asked in your question was prefaced with a statement that the only way to stabilize carbon dioxide emissions was with a tax. And my response was, and is, that the Bush-Quayle administration has itself put out a lengthy and comprehensive study showing that that is not necessary.

Now, as to what I have proposed in my book, I believe that we ought to reduce taxes on things that we need more of, like work. And I think that we ought to shift the burden away from work and toward pollution. If we have incentives for more work and disincentives for more pollution, I think that is a rational direction in which to proceed in the future.

Senator MCCONNELL. So the answer was yes that you did advocate a carbon tax in your book and in your legislation.

Senator GORE. You know, I am surprised you have not read it yet, Senator McConnell.

Senator MCCONNELL. Could you just answer the question; did you or did you not advocate a carbon tax in your book?

Senator GORE. I have proposed a combination of measures, including a reduction of taxes on work, a reduction in payroll taxes, to be replaced by taxes on pollution, including carbon dioxide.

Senator MCCONNELL. You are saying read my lips; no new carbon taxes?

INTERGOVERNMENTAL PANEL ON CLIMATE CHANGE

SECOND ASSESSMENT REPORT

The IPCC undertook its Second Assessment Report in a very different geopolitical context than it did its First Assessment Report in 1990. With the end of the Cold War and the advent of the United Nations Framework Convention on Climate Change, climate change politics had by 1995 become more prominent, more pressing, and more controversial. Criticisms of the IPCC process came from both developing-world nations interested in representation and new scientific resources and developed-world conservatives hoping to avoid binding emissions commitments. These criticisms focused in part on the IPCC's method of determining consensus, prompting the organization to clarify the assessment process—which now included peer review—in its executive summary. How does the IPCC try in this excerpt to answer these criticisms? How convincing are the responses? Despite scientists' best efforts, the report again drew heavy fire from climate change skeptics and conservative pundits, some of whom lobbed charges of conspiracy at IPCC authors, most prominently chapter 8 lead author Ben Santer.* In addition to its acknowledgment of the need for transparency, the Second Assessment Report (SAR) self-consciously spoke to the objectives of the UNFCCC and the sustainable development paradigm it

---

IPCC SAR SYR, *Climate Change 1995: A Report of the Intergovernmental Panel on Climate Change*, Second Assessment Report of the Intergovernmental Panel on Climate Change, 1996. Excerpt: pp. v, vii–viii, 3–4, 21–22.

*Naomi Oreskes and Erik Conway, "The Denial of Global Warming," in *Merchants of Doubt: How a Handful of Scientists Obscured the Truth on Issues from Tobacco Smoke to Global Warming* (New York: Bloomsbury Press, 2010), 169–215.

espoused. As a result, Working Group 2 on climate change impacts and Working Group 3 on policy responses became more prominent, sharing equal billing with Working Group 1 on basic science. Working Group 1's report addressed a number of new findings and directions in research—detecting the signal of climate change and understanding the effects of aerosols were some of the more important issues—but if anything SAR's statement that the "balance of evidence suggests a discernible human influence on climate" soft-pedaled the consensus on climate change, even as Working Groups 2 and 3 pointed toward potential paths forward on responding to the problem. How does the synthesis report, excerpted here, reflect these various scientific and political developments?

· ·

PREFACE

The Intergovernmental Panel on Climate Change (IPCC) was jointly established by the World Meteorological Organization and the United Nations Environment Programme in 1988, in order to: (i) assess available scientific information on climate change, (ii) assess the environmental and socio-economic impacts of climate change, and (iii) formulate response strategies. The IPCC First Assessment Report was completed in August 1990 and served as the basis for negotiating the UN Framework Convention on Climate Change....

In 1992, the Panel reorganized its Working Groups II and III to assess, respectively, the impacts and response options, and the social and economic aspects of climate change. It committed itself to completing its Second Assessment in 1995, not only updating the information on the same range of topics as in the First Assessment, but also including the new subject area of technical issues related to the socio-economic aspects of climate change....

We take this opportunity, because of much misinformation and misunderstanding on the subject, to inform the reader on how the IPCC conducts its assessments.

1. The Panel at the outset decides the content, broken down into chapters, of the report of each of its Working Groups. A writing team of three to six experts (on some rare occasions, more) is constituted for the initial drafting and subsequent revisions of a chapter. Governments and intergovernmental and non-governmental organizations are requested to nominate individuals with appropriate expertise for consideration for inclusion in the writing teams. The publication record of the nominees and other relevant information are also requested. Lists of such individuals are compiled from which the writing team is selected by the Bureau of the Working Group concerned (i.e., the Co-Chairmen and the Vice-Chairmen of the Working Group). The IPCC requires that at least one member of each writing team be from the developing world.

2. The reports are required to have a Summary for Policymakers (SPM). The SPM should reflect the state-of-the-art understanding of the subject matter and be written in a manner that is readily comprehensible to the non-specialist. Differing but scientifically or technically well-founded views should be so exposed in the reports and the SPMs, if they cannot be reconciled in the course of the assessment.

3. The writing teams draft the chapters and the material for inclusion in the SPMs. The drafts are based on literature published in peer-reviewed journals and reports of professional organizations such as the International Council of Scientific Unions, the World Meteorological Organization, the United Nations Environment Programme, the World Health Organization and the United Nations Food and Agriculture Organization. Sometimes, the IPCC holds workshops to collect information that is otherwise not readily available; this is particularly done to encourage information-gathering on and in the developing countries.

4. Each draft chapter is sent to tens of experts worldwide for expert review. The reviewers are also chosen from nominations made by governments and organizations. The mandated time for this review is six weeks. The draft, revised in the light of the comments received, is sent to governments and organizations for their technical review. The mandated time for this (second) review is also six weeks. In some cases, the

expert and government reviews are conducted simultaneously when the time factor would not permit sequential reviews.

5. The draft is revised a second time in the light of the reviews received from governments and organizations. It is then sent to governments (and organizations) one month in advance of the session of the Working Group which would consider it. The Working Group approves the SPM line by line and accepts the underlying chapters; the two together constitute the Report of the Working Group. It is not practical for the Working Group to approve its Report which usually runs to two hundred pages or more. The meaning of the term acceptance in this context is that the underlying chapters and the SPM are consistent with each other.

6. When the Working Group approves the SPM, selected members of the writing teams—from the developing as well as the developed worlds—are present and the text of the SPM is revised at the session with their concurrence. Thus, in reality, the Reports of the Working Groups are written and revised by experts and reviewed by other experts.

7. The Report of the Working Group (with the approved SPM) is sent to governments and organizations one month before the session of the IPCC which would consider it for acceptance.

8. The reader may note that the IPCC is a fully intergovernmental, scientific-technical body. All States that are Members of the United Nations and of the World Meteorological Organization are Members of the IPCC and its Working Groups. As such, governments approve the SPMs and accept the underlying chapters, which are, as stated earlier, written and revised by experts.

May we reiterate that the reports of the IPCC and of its Working Groups contain the factual basis of the issue of climate change, gleaned from available expert literature and further carefully reviewed by experts and governments. In total more than two thousand experts worldwide participate in drafting and reviewing them.

Governments of the world approve/accept them for their scientific-technical content. The final product is written by experts selected worldwide and accepted by governments sitting in plenary sessions. . . .

ADDRESSING THE UNFCCC ARTICLE 2

1.1 Following a resolution of the Executive Council of the World Meteorological Organization (July 1992), the IPCC decided to include an examination of approaches to Article 2, the Objective of the UN Framework Convention on Climate Change (UNFCCC), in its work programme. . . .

The ultimate objective of the UNFCCC, as expressed in Article 2 is:

> " . . . stabilization of greenhouse gas concentrations in the atmosphere at a level that would prevent dangerous anthropogenic interference with the climate system. Such a level should be achieved within a timeframe sufficient to allow ecosystems to adapt naturally to climate change, to ensure that food production is not threatened and to enable economic development to proceed in a sustainable manner."

1.5 The challenges presented to the policymaker by Article 2 are the determination of what concentrations of greenhouse gases might be regarded as "dangerous anthropogenic interference with the climate system" and the charting of a future which allows for economic development which is sustainable. The purpose of this synthesis report is to provide scientific, technical and socio-economic information that can be used, inter alia, in addressing these challenges. . . .

1.7 Given current trends of increasing emissions of most greenhouse gases, atmospheric concentrations of these gases will increase through the next century and beyond. With the growth in atmospheric concentrations of greenhouse gases, interference with the climate system will grow in magnitude and the likelihood of adverse impacts from climate change that could be judged dangerous will become greater. Therefore, possible pathways of future net emissions were considered which might lead to stabilization at different levels and the general constraints these imply. This consideration forms the next part of the report and is followed by a summary of the technical and policy options for reducing emissions and enhancing sinks of greenhouse gases.

1.8 The report then addresses issues related to equity and to ensuring that economic development proceeds in a sustainable manner. This involves addressing, for instance, estimates of the likely damage of climate change impacts, and the impacts, including costs and benefits, of adaptation and mitigation. Finally, a number of insights from available studies point to ways of taking initial actions (see the section on Road Forward) even if, at present, it is difficult to decide upon a target for atmospheric concentrations, including considerations of time-frames, that would prevent "dangerous anthropogenic interference with the climate system."

1.9 Climate change presents the decision maker with a set of formidable complications: considerable remaining uncertainties inherent in the complexity of the problem, the potential for irreversible damages or costs, a very long planning horizon, long time lags between emissions and effects, wide regional variations in causes and effects, an irreducibly global problem, and a multiple of greenhouse gases and aerosols to consider. Yet another complication is that effective protection of the climate system requires international cooperation in the context of wide variations in income levels, flexibility and expectations of the future; this raises issues of efficiency and intra-national, international and intergenerational equity. Equity is an important element for legitimizing decisions and promoting cooperation....

SUMMARY FOR POLICYMAKERS: THE SCIENCE OF CLIMATE CHANGE

Considerable progress has been made in the understanding of climate change science since 1990 and new data and analyses have become available.

1. Greenhouse Gas Concentrations Have Continued to Increase

Increases in greenhouse gas concentrations since pre-industrial times (i.e., since about 1750) have led to a positive radiative forcing of climate, tending to warm the surface and to produce other changes of climate....

2. Anthropogenic Aerosols Tend to Produce Negative Radiative Forcings

- Tropospheric aerosols (microscopic airborne particles) resulting from combustion of fossil fuels, biomass burning and other sources have led to a negative direct forcing of about 0.5 Wm^{-2}, as a global average, and possibly also to a negative indirect forcing of a similar magnitude. While the negative forcing is focused in particular regions and subcontinental areas, it can have continental to hemispheric scale effects on climate patterns.
- Locally, the aerosol forcing can be large enough to more than offset the positive forcing due to greenhouse gases.
- In contrast to the long-lived greenhouse gases, anthropogenic aerosols are very short-lived in the atmosphere, hence their radiative forcing adjusts rapidly to increases or decreases in emissions.

3. Climate Has Changed over the Past Century

At any one location, year-to-year variations in weather can be large, but analyses of meteorological and other data over large areas and over periods of decades or more have provided evidence for some important systematic changes.

- Global mean surface air temperature has increased by between about 0.3 and 0.6°C since the late 19th century; the additional data available since 1990 and the re-analyses since then have not significantly changed this range of estimated increase.
- Recent years have been among the warmest since 1860, i.e., in the period of instrumental record, despite the cooling effect of the 1991 Mt Pinatubo volcanic eruption.
- Global sea level has risen by between 10 and 25 cm over the past 100 years and much of the rise may be related to the increase in global mean temperature....

4. The Balance of Evidence Suggests a Discernible Human Influence on Global Climate

Since the 1990 IPCC Report, considerable progress has been made in attempts to distinguish between natural and anthropogenic influences on climate. . . .

The most important results related to the issues of detection and attribution are:

- The limited available evidence from proxy climate indicators suggests that the 20th century global mean temperature is at least as warm as any other century since at least 1400 A.D. . . .
- More convincing recent evidence for the attribution of a human effect on climate is emerging from pattern-based studies, in which the modeled climate response to combined forcing by greenhouse gases and anthropogenic sulphate aerosols is compared with observed geographical, seasonal and vertical patterns of atmospheric temperature change. These studies show that such pattern correspondences increase with time, as one would expect, as an anthropogenic signal increases in strength. Furthermore, the probability is very low that these correspondences could occur by chance as a result of natural internal variability only. The vertical patterns of change are also inconsistent with those expected for solar and volcanic forcing.
- Our ability to quantify the human influence on global climate is currently limited because the expected signal is still emerging from the noise of natural variability, and because there are uncertainties in key factors. . . . Nevertheless, the balance of evidence suggests that there is a discernible human influence on global climate.

UNITED NATIONS

THE KYOTO PROTOCOL TO THE UNITED NATIONS FRAMEWORK CONVENTION ON CLIMATE CHANGE

The debate over the Kyoto Protocol unfolded as a continuation of the debate over the ratification of the UNFCCC. In 1997, however, the institutional roles were reversed; whereas a Democratic Congress criticized a reluctant Republican president for his cautious approach to the UNFCCC in 1992, by 1997 a Republican Congress stood in the way of a Democratic president's more aggressive efforts to craft and ratify the Kyoto Protocol. The protocol's language echoed that of the UNFCCC, but three new developments reflected a U.S.-led effort to incorporate flexible market solutions into climate change governance. First, a process called "joint implementation" enabled Annex I and Annex II countries to work together on mutually beneficial projects to reduce overall emissions. This was particularly targeted at helping reform "economies in transition" from communism to capitalism in the years following the fall of the Soviet Union. Second, the "clean development mechanism" allowed developed world countries (Annex I) to take credit for funding and implementing greenhouse gas reductions in developing countries. And finally, emissions trading (also called cap-and-trade) created a compliance market for carbon credits that enabled countries to pursue least-cost greenhouse gas reductions across the

---

Kyoto Protocol to the United Nations Framework Convention on Climate Change, December 10, 1997, U.N. Doc FCCC/CP/1997/7/Add.1, 37 I.L.M. 22 (1998). Excerpt: pp. 1–2, 3–4, 5, 6–7, 9–10, 11–12, 15, 19, 20. Copyright 1998 United Nations. Reprinted with the permission of the United Nations.

board, either through joint implementation and the clean development mechanism or through market-based emissions trading. The whole system relied on quantifications of greenhouse gas emissions, now measured in units defined by the equivalent greenhouse equivalent of a single unit of CO_2, facilitated by methodologies laid out by the IPCC. Given Bush's "no regrets" strategy, why might the United States, more than any other nation, have pushed these "flexibility mechanisms" as a way to achieve least-cost emissions reductions based on market force? Despite U.S. delegation chief Stuart Eizenstat's success at brokering a deal on flexibility mechanisms (with help from Vice President Gore), the Kyoto Protocol still failed to satisfy domestic conservatives, who worried about the lack of emissions responsibilities in the developing world, among other things.

· ·

The Parties to this Protocol,

Being Parties to the United Nations Framework Convention on Climate Change,

hereinafter referred to as "the Convention,"

In pursuit of the ultimate objective of the Convention as stated in its Article 2,

Recalling the provisions of the Convention,

Being guided by Article 3 of the Convention,

Pursuant to the Berlin Mandate adopted by decision 1/CP.1 of the Conference of the Parties to the Convention at its first session,

Have agreed as follows:

ARTICLE 1

For the purposes of this Protocol, the definitions contained in Article 1 of the Convention shall apply. In addition:

1. "Conference of the Parties" means the Conference of the Parties to the Convention.
2. "Convention" means the United Nations Framework Convention on Climate Change, adopted in New York on 9 May 1992.

ARTICLE 2

1. Each Party included in Annex I, in achieving its quantified emission limitation and reduction commitments under Article 3, in order to promote sustainable development, shall:
   (a) Implement and/or further elaborate policies and measures in accordance with its national circumstances, such as:
      (i) Enhancement of energy efficiency in relevant sectors of the national economy;
      (ii) Protection and enhancement of sinks and reservoirs of greenhouse gases not controlled by the Montreal Protocol, taking into account its commitments under relevant international environmental agreements; promotion of sustainable forest management practices, afforestation and reforestation;
      (iii) Promotion of sustainable forms of agriculture in light of climate change considerations;
      (iv) Research on, and promotion, development and increased use of, new and renewable forms of energy, of carbon dioxide sequestration technologies and of advanced and innovative environmentally sound technologies;
      (v) Progressive reduction or phasing out of market imperfections, fiscal incentives, tax and duty exemptions and subsidies in all greenhouse gas emitting sectors that run counter to the objective of the Convention and application of market instruments;
      (vi) Encouragement of appropriate reforms in relevant sectors aimed at promoting policies and measures which limit or reduce emissions of greenhouse gases not controlled by the Montreal Protocol;
      (vii) Measures to limit and/or reduce emissions of greenhouse gases not controlled by the Montreal Protocol in the transport sector;
      (viii) Limitation and/or reduction of methane emissions through recovery and use in waste management, as well as in the production, transport and distribution of energy;

(b) Cooperate with other such Parties to enhance the individual and combined effectiveness of their policies and measures adopted under this Article, pursuant to Article 4, paragraph 2 (e) (i), of the Convention. To this end, these Parties shall take steps to share their experience and exchange information on such policies and measures, including developing ways of improving their comparability, transparency and effectiveness. The Conference of the Parties serving as the meeting of the Parties to this Protocol shall, at its first session or as soon as practicable thereafter, consider ways to facilitate such cooperation, taking into account all relevant information. . . .

ARTICLE 3

1. The Parties included in Annex I shall, individually or jointly, ensure that their aggregate anthropogenic carbon dioxide equivalent emissions of the greenhouse gases listed in Annex A do not exceed their assigned amounts, calculated pursuant to their quantified emission limitation and reduction commitments inscribed in Annex B and in accordance with the provisions of this Article, with a view to reducing their overall emissions of such gases by at least 5 per cent below 1990 levels in the commitment period 2008 to 2012.
2. Each Party included in Annex I shall, by 2005, have made demonstrable progress in achieving its commitments under this Protocol.
3. The net changes in greenhouse gas emissions by sources and removals by sinks resulting from direct human-induced land-use change and forestry activities, limited to afforestation, reforestation and deforestation since 1990, measured as verifiable changes in carbon stocks in each commitment period, shall be used to meet the commitments under this Article of each Party included in Annex I. . . .
4. Prior to the first session of the Conference of the Parties serving as the meeting of the Parties to this Protocol, each Party included in Annex I shall provide, for consideration by the Subsidiary Body for Scientific and Technological Advice, data to establish its level of carbon stocks in 1990 and to enable an estimate to be made of its changes in carbon stocks in subsequent years. . . .

ARTICLE 4

1. Any Parties included in Annex I that have reached an agreement to fulfill their commitments under Article 3 jointly, shall be deemed to have met those commitments provided that their total combined aggregate anthropogenic carbon dioxide equivalent emissions of the greenhouse gases listed in Annex A do not exceed their assigned amounts calculated pursuant to their quantified emission limitation and reduction commitments inscribed in Annex B and in accordance with the provisions of Article 3. The respective emission level allocated to each of the Parties to the agreement shall be set out in that agreement....

ARTICLE 6

1. For the purpose of meeting its commitments under Article 3, any Party included in Annex I may transfer to, or acquire from, any other such Party emission reduction units resulting from projects aimed at reducing anthropogenic emissions by sources or enhancing anthropogenic removals by sinks of greenhouse gases in any sector of the economy, provided that:
    (a) Any such project has the approval of the Parties involved;
    (b) Any such project provides a reduction in emissions by sources, or an enhancement of removals by sinks, that is additional to any that would otherwise occur;
    (c) It does not acquire any emission reduction units if it is not in compliance with its obligations under Articles 5 and 7; and
    (d) The acquisition of emission reduction units shall be supplemental to domestic actions for the purposes of meeting commitments under Article 3....

ARTICLE 10

All Parties, taking into account their common but differentiated responsibilities and their specific national and regional development priorities, objectives and circumstances, without introducing any new commitments for Parties not included in Annex I, but reaffirming existing commitments under Article 4, paragraph 1, of the Convention, and continuing to advance the implementation of these commitments in order to achieve sustainable development, taking into account Article 4, paragraphs 3, 5 and 7, of the Convention, shall:

(a) Formulate, where relevant and to the extent possible, cost-effective national and, where appropriate, regional programmes to improve the quality of local emission factors, activity data and/or models which reflect the socio-economic conditions of each Party for the preparation and periodic updating of national inventories of anthropogenic emissions by sources and removals by sinks of all greenhouse gases not controlled by the Montreal Protocol, using comparable methodologies to be agreed upon by the Conference of the Parties, and consistent with the guidelines for the preparation of national communications adopted by the Conference of the Parties; ...

(c) Cooperate in the promotion of effective modalities for the development, application and diffusion of, and take all practicable steps to promote, facilitate and finance, as appropriate, the transfer of, or access to, environmentally sound technologies, know-how, practices and processes pertinent to climate change, in particular to developing countries, including the formulation of policies and programmes for the effective transfer of environmentally sound technologies that are publicly owned or in the public domain and the creation of an enabling environment for the private sector, to promote and enhance the transfer of, and access to, environmentally sound technologies; ...

ARTICLE 12

1. A clean development mechanism is hereby defined.
2. The purpose of the clean development mechanism shall be to assist Parties not included in Annex I in achieving sustainable development and in contributing to the ultimate objective of the Convention, and to assist Parties included in Annex I in achieving compliance with their quantified emission limitation and reduction commitments under Article 3.

3. Under the clean development mechanism:
   (a) Parties not included in Annex I will benefit from project activities resulting in certified emission reductions; and
   (b) Parties included in Annex I may use the certified emission reductions accruing from such project activities to contribute to compliance with part of their quantified emission limitation and reduction commitments under Article 3, as determined by the Conference of the Parties serving as the meeting of the Parties to this Protocol. . . .

5. Emission reductions resulting from each project activity shall be certified by operational entities to be designated by the Conference of the Parties serving as the meeting of the Parties to this Protocol, on the basis of:
   (a) Voluntary participation approved by each Party involved;
   (b) Real, measurable, and long-term benefits related to the mitigation of climate change; and
   (c) Reductions in emissions that are additional to any that would occur in the absence of the certified project activity. . . .

ARTICLE 18

The Conference of the Parties serving as the meeting of the Parties to this Protocol shall, at its first session, approve appropriate and effective procedures and mechanisms to determine and to address cases of non-compliance with the provisions of this Protocol, including through the development of an indicative list of consequences, taking into account the cause, type, degree and frequency of non-compliance. Any procedures and mechanisms under this Article entailing binding consequences shall be adopted by means of an amendment to this Protocol. . . .

ANNEX A

Greenhouse Gases

Carbon dioxide (CO_2)
Methane (CH_4)
Nitrous oxide (N_2O)
Hydrofluorocarbons (HFCs)
Perfluorocarbons (PFCs)
Sulphur hexafluoride (SF_6) ...

ANNEX B

Party	Quantified Emissions Limitation or Reduction Commitment (percentage of base year or period)
Australia	108
Austria	92
Belgium	92
Bulgaria*	92
Canada	94
Croatia*	95
Czech Republic*	92
Denmark	92
Estonia*	92
European Community	92
Finland	92
France	92
Germany	92
Greece	92
Hungary*	94
Iceland	110
Ireland	92
Italy	92
Japan	94
Latvia*	92

Party	Quantified Emissions Limitation or Reduction Commitment (percentage of base year or period)
Liechtenstein	92
Lithuania*	92
Luxembourg	92
Monaco	92
Netherlands	92
New Zealand	100
Norway	101
Poland*	94
Portugal	92
Romania*	92
Russian Federation*	100
Slovakia*	92
Slovenia*	92
Spain	92
Sweden	92
Switzerland	92
Ukraine*	100
United Kingdom of Great Britain and Northern Ireland	92
United States of America	93

* Countries that are undergoing the process of transition to a market economy.

THE BYRD-HAGEL RESOLUTION

As with any treaty or treaty protocol, President Clinton would need the consent of the Senate for American participation in the Kyoto Protocol once it was written. In 1997, however, the Senate looked with suspicion at binding commitments to specific emissions reductions on well-defined timetables. In the Senate, both Democratic and Republican senators worried about the trade impacts of a treaty protocol that would require significant emissions reductions in the United States but would require nothing from trade partners from the developing world—including an increasingly competitive China. In 1997, Democratic senator Robert Byrd of West Virginia—a coal state—and Republican Chuck Hagel of Nebraska drafted a resolution that outlined the Senate's concerns about developments at Kyoto. The Byrd-Hagel Resolution was not a vote on the Kyoto Protocol; rather, the resolution conveyed the "sense of the Senate" that it would not support an international emissions protocol unless it included the developing world and could be demonstrated not to harm the U.S. economy. The Senate passed the Byrd-Hagel Resolution 95–0, and though President Clinton signed the Kyoto Protocol, neither he nor his successor ever sent it to the Senate for ratification. Why would the Senate ratify the IPCC in 1993 and then come out so strongly against the Kyoto Protocol in 1997?

• •

Byrd-Hagel Resolution, S. Res. 98, 105th Cong. (1997). Excerpt: pp. 1–5.

RESOLUTION

Expressing the sense of the Senate regarding the conditions for the United States becoming a signatory to any international agreement on greenhouse gas emissions under the United Nations Framework Convention on Climate Change.

Whereas the United Nations Framework Convention on Climate Change (in this resolution referred to as the "Convention"), adopted in May 1992, entered into force in 1994 and is not yet fully implemented;

Whereas the Convention, intended to address climate change on a global basis, identifies the former Soviet Union and the countries of Eastern Europe and the Organization For Economic Co-operation and Development (OECD), including the United States, as "Annex I Parties," and the remaining 129 countries, including China, Mexico, India, Brazil, and South Korea, as "Developing Country Parties";

Whereas in April 1995, the Convention's "Conference of the Parties" adopted the so-called "Berlin Mandate";

Whereas the "Berlin Mandate" calls for the adoption, as soon as December 1997, in Kyoto, Japan, of a protocol or another legal instrument that strengthens commitments to limit greenhouse gas emissions by Annex I Parties for the post-2000 period and establishes a negotiation process called the "Ad Hoc Group on the Berlin Mandate";

Whereas the "Berlin Mandate" specifically exempts all Developing Country Parties from any new commitments in such negotiation process for the post-2000 period;

Whereas although the Convention, approved by the United States Senate, called on all signatory parties to adopt policies and programs aimed at limiting their greenhouse gas (GHG) emissions, in July 1996 the Undersecretary of State for Global Affairs called for the first time for "legally binding" emission limitation targets and timetables for Annex I Parties, a position reiterated by the Secretary of State in testimony before the Committee on Foreign Relations of the Senate on January 8, 1997;

Whereas greenhouse gas emissions of Developing Country Parties are rapidly increasing and are expected to surpass emissions of the United States and other OECD countries as early as 2015;

Whereas the Department of State has declared that it is critical for the Parties to the Convention to include Developing Country Parties in the next steps for global action and, therefore, has proposed that consideration of additional steps to include limitations on Developing Country Parties' greenhouse gas emissions would not begin until after a protocol or other legal instrument is adopted in Kyoto, Japan in December 1997;

Whereas the exemption for Developing Country Parties is inconsistent with the need for global action on climate change and is environmentally flawed;

Whereas the Senate strongly believes that the proposals under negotiation, because of the disparity of treatment between Annex I Parties and Developing Countries and the level of required emission reductions, could result in serious harm to the United States economy, including significant job loss, trade disadvantages, increased energy and consumer costs, or any combination thereof; and

Whereas it is desirable that a bipartisan group of Senators be appointed by the Majority and Minority Leaders of the Senate for the purpose of monitoring the status of negotiations on Global Climate Change and reporting periodically to the Senate on those negotiations: Now, therefore, be it

Resolved, That it is the sense of the Senate that—

(1) the United States should not be a signatory to any protocol to, or other agreement regarding, the United Nations Framework Convention on Climate Change of 1992, at negotiations in Kyoto in December 1997, or thereafter, which would—

(A) mandate new commitments to limit or reduce greenhouse gas emissions for the Annex I Parties, unless the protocol or other agreement also mandates new specific scheduled commitments to limit or reduce greenhouse gas emissions for Developing Country Parties within the same compliance period, or

(B) would result in serious harm to the economy of the United States; and

(2) any such protocol or other agreement which would require the advice and consent of the Senate to ratification should be accompanied by a detailed explanation of any legislation or regulatory actions that may be required to implement the protocol or other agreement and should also be accompanied by an analysis of

the detailed financial costs and other impacts on the economy of the United States which would be incurred by the implementation of the protocol or other agreement.

Sec. 2. The Secretary of the Senate shall transmit a copy of this resolution to the President.

PART 6

THE PAST, THE PRESENT, AND THE FUTURE

So far, the bulk of this collection has focused on the way scientists and other experts have worked to incorporate information about climate change into the institutional frameworks of governments, scientific and environmental groups, and international organizations like the United Nations. In part, this narrow focus on experts is an artifact of the history of climate change discourse itself. Between the early exploration of physicists like Fourier and Tyndall in the nineteenth century and the advent of the UNFCCC in 1992, the most important developments in the history of global warming have involved a difficult process of translation, wherein scientists have struggled to make their particular forms of knowledge meaningful to people who make policy and set scientific and environmental priorities. Periodically, these efforts had a public face; Stephen Schneider wrote his 1976 *The Genesis Strategy* for a popular audience, for example. But until the late 1980s, even scientists' public forays had more to do with showing what climate change *is* and what institutions might do about it than they did with exploring what climate change *means* in a broader cultural context.

Since the late 1980s, that story has changed. In the 1990s, conversations about global warming came to include a much wider variety of experts—as well as a host of authors working outside of their expertise—and they have occurred in a multitude of new forums and in new forms of media. The effort to collect and disseminate expert knowledge on climate change and turn that knowledge into policy certainly continued throughout the 1990s and early 2000s, both in the ongoing work of the IPCC and domestically in a number of failed pieces of congressional legislation, most notably the American Clean Energy and Security Act of 2009. But while the science-policy interface remained important—and perhaps still paramount—in the 2000s, making sense of the more recent history of climate change requires a broadening of the net in a way that previous periods do not. Here, I present a subset of documents from the expanding literature on climate change that demonstrate some of the ways experts and luminaries from *outside* of the core disciplines of climate science have tried to make sense of what climate change means for their particular intellectual, religious, ideological, or practical approaches to the world.

Bill McKibben's 1989 book, *The End of Nature*, provides a starting point for the expansion of modern global warming discourse. McKibben married an exposé of what scientists then knew about global warming with a thoughtful and pointed exploration of what that knowledge meant for humans' relationship to nature in the coming decade. A successful author, journalist, and creative writing professor, he sought to frame global warming not just as a technical and political challenge, but also as a real threat to some of the nation's cherished moral and political foundations. *The End of Nature* stands among the most widely read works of environmental literature of the global warming era, and more than any other book, it helped to define climate change for a whole generation of Americans. How does McKibbens' text differ from the scientific, governmental, and even "popular" presentations of climate change from previous decades? What exactly has changed for McKibben that is worth lamenting?

McKibben's basic insight—that beyond its immediate physical manifestations, global warming has somehow made the world we inhabit fundamentally different—underscores the common theme that unites the documents from this chapter. When you look at each of these documents, it is worth asking, for the authors, what has climate change changed? For a subgroup of geologists, the question revolved (and still revolves) around a fundamental change in the way modern humans have begun to affect the geophysical processes of the earth, a change that Paul Crutzen and Eugene Stoermer argue requires a new geological distinction for the current epoch: the "anthropocene." For Ted Nordhaus and Michael Schellenberger, climate change changed the way we ought to think about environmental politics in the early 2000s, while for the Supreme Court, climate change challenged the limits of the concept of standing and the rulemaking authority of executive agencies like the EPA. For economist Nicholas Stern and his critics, climate change required a reconsideration of the way we understand the responsibilities of current generations for their collective children and grandchildren, coded in an economic term, the "discount rate."[1] And for Pope Francis, whose 2015 encyclical concludes this collection, finding meaning in climate change requires that we rethink nothing less than our relationships to the earth, to each other, and to God.

The question that each of these authors address—what does climate change change?—is at its heart a historical one. Throughout the history of climate change, scientists, environmentalists, policymakers, and diplomats have worked to translate the technical, future-tense problem of climate change into the language of present policy and action. Since the late 1980s, however, nonscientists like McKibben, Chief Justice Roberts, and Pope Francis have attempted to go beyond that process of translation, taking the reality of climate change as a starting point and using it to link the past, present, and future in an effort to make meaning out of the reality of a warming world. In the same way that making climate change history requires us to question what looks familiar and what looks strange about documents from global warming's past, these authors have constructed stories about the era of climate change with a "before" and an "after" that highlight continuity and change. And yet, neither Supreme Court decisions nor papal encyclicals are self-consciously "histories," despite the claims they make to the past. How do we approach documents from global warming's history that use the past to make sense of the climatic present?

One place to start is by thinking about how an author or a document draws on the past as a source of authority. Take, for example, two documents that at first glance appear very different: the Supreme Court's 2007 decision in *Massachusetts v. EPA* and Pope Francis's 2015 encyclical, *Laudato Si': On Care for Our Common Home*. The 2007 court decision marked the end of a four-year process of litigation in which Massachusetts, along with eleven other states and numerous environmental organizations, sued the Environmental Protection Agency to begin regulating greenhouse gas emissions through fuel efficiency standards under the 1990 Clean Air Act. The main issues of the case involved "standing," or who can claim harm from climate change sufficient to have grounds for bringing suit in a court of law; and statutory intent, here the question of what exactly the U.S. Congress meant the Clean Air Act to cover when it passed the legislation in 1990. By contrast, the 2015 papal encyclical, circulated eight years later, addressed climatic and environmental change not from a legal perspective but from a spiritual one. An encyclical is a document concerning Catholic doctrine distributed among official church bishops and all those in communion with the Holy See, and *Laudato Si'* was the relatively new pope's second (the first was drafted primarily by his predecessor). While the intended audiences and purposes of the two documents are obviously quite different, the pope and the Supreme Court use the past to make sense of their institutions' present position on climate-related issues in remarkably similar ways. What formal elements do the court's decision and the encyclical share that help them make sense of the present in terms of the past? Where do the pope and the court turn for authority, respectively? What roles do previous papal and Supreme Court documents play in the way the nation's highest court and Catholicism's highest priest work climate change into their institutions' missions? I have included the notation apparatus of these two documents here to highlight the formal appeals to precedent in the two documents.

The formal similarities between the 2015 encyclical and *Massachusetts v. EPA* present a fairly straightforward example of how authors use precedent as a guide to grappling with climate change. But authors like McKibben and scientists like Crutzen and Stoermer appeal to precedent in less clearly codified ways as well, especially where climate change represents a break with that precedent. For Crutzen and Stoermer, what is unprecedented about anthropogenic climate change, and why is that break with precedent important enough to require a new nomenclature? Again, I have included the notes. How and to what extent do Crutzen and Stoermer's notes contribute to *their* argument? What is at stake in identifying 1784 as a starting point for the anthropocene? How do their particular concerns with precedent differ from McKibben's concerns? Where does McKibben turn for authority in *The End of Nature*, and how do his appeals to the past differ from those of Pope Francis or Chief Justice Roberts?

If you have read this book carefully, these uses of precedent to link the past to the present may raise some red flags about what I have called throughout this work the "presentist paradox." And rightly so. The presentist paradox arises from the seemingly contradictory impulses to study history that is relevant to a present issue—climate change—and to take history on its own terms, avoiding the distorting lens of our present interest. In truth, however, these are not necessarily incommensurate impulses. Each of the documents here takes the past as a standard against which to compare the present and the future in a warming world, and can thus be called "presentist." And yet, where the authors engage critically with history, that comparison serves as a great asset in making meaning out of climate change.

This, of course, is an argument for history. At its best, the appeal to precedent for guidance on global warming can focus our attention on the way climate change has begun to modify the ecological, political, legal, and social contexts in which we make decisions and interpret our lives. In *Massachusetts v. EPA*, for example, the court critically examined the context of the passage of the Clean Air Act of 1990 to determine whether regulating CO_2 fit within the intent of the law at the time it was passed. Both the majority and the minority recognized that fighting global warming was not the primary focus of the 1990 bill—that is, they took the bill on its own historical terms—but ultimately the majority concluded that given new information, it made sense to adapt a statute from the past to address a related problem of the present. Nordhaus and Shellenberger, meanwhile, recognize that the litigation and lobbying which worked so well on local pollution issues of the 1970s no longer provide a good model for environmental advocacy, largely because the context of environmental politics has changed. A strong historical sensibility enables lawmakers, political strategists, and others to recognize the way the changing contexts of climate change require us to adjust our approaches to the problem over time. To borrow from Nordhaus and Shellenberger, contextually rich historical thinking helps us prevent "fighting the last war."

The utility of thinking historically about climate change reaches beyond political strategy, however. As the documents in this chapter demonstrate, history enables us to make judgments about the dangers and opportunities of climate change by highlighting those things about the past and present that we value in the face of a warming world. Critical historical thinking prompts us to think more deeply about these judgments and how we make them. As McKibben's work highlights, many of our collective concerns over climate change represent a sort of lament. *The End of Nature* uses the past as a normative foil for the present, positing a simpler, purer time when a simpler, purer place called nature existed outside the realm of human influence. But his framing is misleading because the static, nonhuman nature to which he appeals is more a historically specific ideal than an absolute location, and the simpler, purer past in which that space exists is a nostalgic fiction. And while McKibben himself has since 1989 become a great advocate for the poor, powerless, and underrepresented communities disproportionately affected by climate change, his ahistorical vision of nature is misleading because it recapitulates ideas that have been deployed to disempower, dispossess, and disenfranchise these very groups for more than a century, helping to make them particularly vulnerable to climate change in the first place.[2] McKibben's presentist framing of the past is thus at once a great asset and a great liability. Overcoming McKibben's problem requires that we think deeply and critically about our historical laments, that we investigate more closely who stands to lose what in the face of a changing climate, and why.

In the chapter introductions throughout this collection, I have introduced a variety of specific strategies for thinking about different types of documents in their historical contexts—tools for using documents to make climate change history. But while making climate change history begins with documents like these, it does not end with documents, and it certainly does not end with *these* documents. (In fact, while history as a discipline tends to focus on documentary evidence, historical thinking—a nuanced attention to context and contingency that helps us take evidence on its own terms—doesn't necessarily require documents at all). Rather, making climate change history culminates in the stories we tell once we have put pieces of evidence in context and made judgments about the winners and losers of our collective laments—in the way we, like the authors included in this final chapter, make meaning out of global warming's past, present, and future more broadly. I have said less about these stories and judgments here because in this context they are not my stories to tell, not my judgments to make. They are yours. Good climate change history remains a vital precondition for meaningful conversations about climate equity, climate justice, and climate solutions. Climate change history needs to be made. I hope this collection provides a helpful starting point for making it.

NOTES

1. See "Background to Stern Review on the Economics of Climate Change," www.hmtreasury.gov.uk/independent_reviews/stern_review_economics_climate_change/sternreview_backgroundtoreview.cfm.
2. For contemporary critiques of McKibben and the ideal of wilderness, see William Cronon, "The Trouble with Wilderness: Or, Getting Back to the Wrong Nature," and Richard White, "Are You an Environmentalist or Do You Work for a Living?" in Cronon, *Uncommon Ground: Toward Reinventing Nature* (New York: W. W. Norton, 1995): 69–90, 171–85.

BILL McKIBBEN

THE END OF NATURE

Bill McKibben's 1989 *The End of Nature* defined the problem of global warming for a generation. Originally serialized by the *New Yorker*, the book introduced global warming for a general audience in the familiar and deeply moralizing terms of a nature writer. The book's publication followed on the heels of James Hansen's highly publicized 1988 testimony on global warming before the Senate Committee on Energy and Natural Resources, and it was the first book on the subject to gain a significant readership outside of a relatively niche science-writing market. *The End of Nature* was less about the science of climate change, however, than about what a world altered by global warming, acid rain, and ozone depletion meant for the cherished concept of nature that Americans appealed to as a locus of solace, morality, and spiritual well-being. As McKibben later wrote, "I was more interested in how [climate change] made me feel—which was more sad than scared. Sad that we had ended the idea that any place was still beyond the human touch."* Here, having explained global warming for the lay reader, McKibben begins his personal exploration of what global warming means. How does nature serve as a form of precedent in this piece? What, exactly, has global warming ruined?

· ·

Bill McKibben, *The End of Nature* (New York: Random House, 1989). Excerpt: pp. 47–49, 58–59, 64.

* Bill McKibben, ed., *The Global Warming Reader: A Century of Writing about Climate Change* (New York: Penguin, 2011), 292.

Almost every day, I hike up the hill out my backdoor. Within a hundred yards the woods swallow me up, and there is nothing to remind me of human society—no trash, no stumps, no fence, not even a real path. Looking out from the high places, you can't see road or house; it is a world apart from man. But once in a while someone will be cutting wood farther down the valley, and the snarl of a chain saw will fill up the woods. It is harder on those days to get caught up in the timeless meaning of the forest, for man is nearby. The sound of the chain saw doesn't blot out the all the noises of the forest or drive the animals away, but it does drive away the feeling that you are in another separate, timeless, wild sphere.

Now that we have changed the most basic forces around us, the noise of that chain saw will always be in the woods. We have changed the atmosphere, and that will change the weather. The temperature and rainfall are no longer to be entirely the work of some separate, uncivilizable force, but instead in part a product of our habits, our economies, our ways of life. Even in the most remote wilderness, where the strictest laws forbid the felling of a single tree, the sound of that saw will be clear, and a walk in the woods will be changed—tainted—by its whine. The world outdoors will mean much the same thing as the world indoors, the hill the same thing as the house.

An idea, a relationship, can go extinct, just like an animal or a plant. The idea in this case is "nature," the separate and wild province, the world apart from man to which he adapted, under whose rules he was born and died. In the past, we spoiled and polluted parts of that nature, inflicted environmental "damage." But that was like stabbing a man with toothpicks: though it hurt, annoyed, degraded, it did not touch vital organs, block the path of the lymph or blood. We never thought that we had wrecked nature. Deep down, we never really thought we could: it was too big and too old; its forces—the wind, the rain, the sun—were too strong, too elemental.

But quite by accident, it turned out that the carbon dioxide and other gases we were producing in our pursuit of a better life—in pursuit of warm houses and eternal economic growth and of agriculture so productive it would free most of us from farming—*could* alter the power of the sun, could increase its heat. And that increase *could* change the patterns of moisture and dryness, breed storms in new places, breed deserts. Those things may or may not have yet begun to happen, but it is too late to altogether prevent them from happening. We have produced the carbon dioxide—we are ending nature.

We have not ended rainfall or sunlight; in fact, rainfall and sunlight may become more important forces in our lives. It is too early to tell exactly how much harder the wind will blow, how much hotter the sun will shine. That is for the future. But the *meaning* of the wind, the sun, the rain—of nature—has already changed. Yes, the wind still blows— but no longer from some other sphere, some inhuman place.

In the summer, my wife and I bike down to the lake nearly every afternoon for a swim. It is a dogleg Adirondack lake, with three beaver lodges, a blue heron, some otter, a family of mergansers, the occasional loon. A few summerhouses cluster at one end, but mostly it is surrounded by wild state land. During the week we swim across and back, a trip of maybe forty minutes—plenty of time to forget everything but the feel of the water around your body and the rippling, muscular joy of a hard kick and the pull of your arms.

But on the weekends, more and more often, someone will bring a boat out for waterskiing, and make pass after pass up and down the lake. And the whole experience changes, changes entirely. Instead of being able to forget everything but yourself, and even yourself except for the muscles and the skin, you must be alert, looking up every dozen strokes to see where the boat is, thinking about what you will do if it comes near. It is not so much the danger—few swimmers, I imagine, ever die by Evinrude. It's not even so much the blue smoke that hangs low over the water. It's that the motorboat gets in your mind. You're forced to think, not feel—to think of human society and of people. The lake is utterly different on these days, just as the planet is utterly different now....

The idea of wildness, in other words, can survive most of the "normal" destruction of nature. Wildness can survive in our minds once the land has been discovered and mapped and even chewed up. It can survive all sorts of pollution, even the ceaseless munching of a million cows. If the ground is dusty and trodden, we look at the sky; if the sky is smoggy, we travel someplace where it's clear; if we can't travel to someplace where it's clear, we imagine ourselves in Alaska or Australia or someplace where it is, and that works nearly as well. Nature, while often fragile in reality, is durable in our imaginations. Wildness, the idea of wildness, has outlasted the exploration of the entire globe. It has endured the pesticides and the pollution. When the nature around us is degraded, we picture it fresh and untainted elsewhere. When elsewhere, too, it rains acid or DDT, we can still imagine that someday soon it will be better, that we will stop polluting and despoiling and instead "restore" nature. (And, indeed, people have begun to do just this sort of work: here in the Adirondacks, helicopters drop huge quantities of lime into lakes in order to reduce their acidity.) In our minds nature suffers from a terrible case of acne, or even skin cancer—but our faith in its essential strength remains, for the damage always seems local.

But now the basis of faith is lost. The idea of nature will not survive the new global pollution—the carbon dioxide and the CFCs and the like. This new rupture with nature is different not only in scope but also in kind from salmon tins in an English stream. We have changed the atmosphere, and thus we are changing the weather. By changing the weather, we make every spot on earth manmade and artificial. We have deprived nature of its independence, and that is fatal to its meaning. Nature's independence *is* its meaning; without it there is nothing but us. . . .

One can, of course, argue that the current crisis, too, is "natural," because man is part of nature. This echoes the views of the earliest Greek philosophers, who saw no difference between matter and consciousness—nature included everything. The British scientist James Lovelock wrote some years ago that "our species with its technology is simply an inevitable part of the natural scene," nothing more than mechanically advanced beavers. In this view, to say that we "ended" nature, or even damaged nature, makes no sense, since we *are* nature, and nothing we can do is "unnatural." This view can be, and is, carried to even greater lengths; Lynn Margulis, for instance, ponders the question of whether robots can be said to be living creatures, since "any invention of human beings is ultimately based on a variety of processes including that of DNA replication, no matter the separation in space or time of that replication from the invention."

But one can argue this forever and still not really feel it. It is a debater's point, a semantic argument. When I say that we have ended nature, I don't mean, obviously, that natural processes have ceased—there is still sunshine and still wind, still growth, still decay. Photosynthesis continues, as does respiration. *But we ended the thing that has, at least in modern times, defined nature for us—its separation from human society.*

PAUL J. CRUTZEN AND EUGENE F. STOERMER

THE ANTHROPOCENE

McKibben's concern over "the end of nature" reflected an important insight into a novel and near-total form of worldwide pollution, but he was not alone in recognizing the change in the scale and character of humanity's impact on the earth. Over the course of the 1990s, scientists in particular worked to catalog the global changes in the atmosphere, the hydrosphere, the biosphere, and the lithosphere that made the world a physically different place than it had been in centuries past. In 2000, atmospheric chemist and Nobel Laureate Paul Crutzen teamed up with biologist Eugene Stoermer to suggest that humanity's impact on the earth's atmosphere was significant enough to constitute a new geological epoch, the "anthropocene." The original Crutzen and Stoermer paper introduced the term in specific reference to the way that CO_2-induced climate change altered geophysical processes on human timescales. Since then, however, the term *anthropocene* has led something of a double life. Among stratigraphers and other geologists, the term has generated significant technical debate about its merit in describing a new epoch (not to be conflated with debate over whether humans have had a global-scale impact on the planet). Outside of that small community, the term has been co-opted to articulate in a different way the insight that McKibben expressed in 1989 about the totality of global change.* Here, the authors introduce the term in the context of the history of industrialization in the original paper, published in the

---

Paul J. Crutzen and Eugene F. Stoermer, "The Anthropocene," *Global Change Newsletter*, no. 41 (May 2000): 17–18.

* See, for example, Dipesh Chakrabarty, "The Climate of History: Four Theses," *Critical Inquiry* 35, no. 2 (January 2009): 197–222; John Robert McNeill,

Global Change Newsletter, a publication of the International Geosphere-Biosphere Programme of the International Council for Science.

• •

The name Holocene ("Recent Whole") for the post-glacial geological epoch of the past ten to twelve thousand years seems to have been proposed for the first time by Sir Charles Lyell in 1833, and adopted by the International Geological Congress in Bologna in 1885.[1] During the Holocene mankind's activities gradually grew into a significant geological, morphological force, as recognized early on by a number of scientists. Thus, G. P. Marsh already in 1864 published a book with the title "Man and Nature," more recently reprinted as "The Earth as Modified by Human Action".[2] Stoppani in 1873 rated mankind's activities as a "new telluric force which in power and universality may be compared to the greater forces of earth" [quoted from Clark].[3] Stoppani already spoke of the anthropozoic era. Mankind has now inhabited or visited almost all places on Earth; he has even set foot on the moon.

The great Russian geologist V. I. Vernadsky[4] in 1926 recognized the increasing power of mankind as part of the biosphere with the following excerpt: "... the direction in which the processes of evolution must proceed, namely towards increasing consciousness and thought, and forms having greater and greater influence on their surroundings." He, the French Jesuit P. Teilhard de Chardin and E. Le Roy in 1924 coined the term "noösphere," the world of thought, to mark the growing role played by mankind's brainpower and technological talents in shaping its own future and environment.

Something New under the Sun: An Environmental History of the Twentieth-Century World (New York: W.W. Norton, 2000).

    1. Encyclopaedia Britannica, Micropaedia, IX (1976).

    2. G. P. Marsh, *The Earth as Modified by Human Action*, Belknap Press, Harvard University Press, 1965.

    3. W. C. Clark, in *Sustainable Development of the Biosphere*, W. C. Clark and R. E. Munn, Eds. (Cambridge University Press, Cambridge, 1986), chapt. 1.

    4. V. I. Vernadski, *The Biosphere*, translated and annotated version from the original of 1926 (Copernicus, Springer, New York, 1998).

The expansion of mankind, both in numbers and per capita exploitation of Earth's resources has been astounding.[5] To give a few examples: During the past 3 centuries human population increased tenfold to 6000 million, accompanied e.g. by a growth in cattle population to 1400 million[6] (about one cow per average size family). Urbanisation has even increased tenfold in the past century. In a few generations mankind is exhausting the fossil fuels that were generated over several hundred million years. The release of SO_2, globally about 160 Tg/year to the atmosphere by coal and oil burning, is at least two times larger than the sum of all natural emissions, occurring mainly as marine dimethyl-sulfide from the oceans;[7] from Vitousek et al.[8] we learn that 30–50% of the land surface has been transformed by human action; more nitrogen is now fixed synthetically and applied as fertilizers in agriculture than fixed naturally in all terrestrial ecosystems; the escape into the atmosphere of NO from fossil fuel and biomass combustion likewise is larger than the natural inputs, giving rise to photochemical ozone ("smog") formation in extensive regions of the world; more than half of all accessible fresh water is used by mankind; human activity has increased the species extinction rate by thousand to ten thousand fold in the tropical rain forests[9] and several climatically important "greenhouse" gases have substantially increased in the atmosphere: CO_2 by more than 30% and CH_4 by even more than 100%. Furthermore, mankind releases many toxic substances in the environment and even some, the chlorofluorocarbon gases, which are not toxic at all, but which nevertheless have led to the Antarctic "ozone hole" and which would have destroyed much of the ozone layer if no international

---

5. B. L. Turner II et al., *The Earth as Transformed by Human Action*, Cambridge University Press, 1990.

6. P. J. Crutzen and T. E. Graedel, in *Sustainable Development of the Biosphere*, W. C. Clark and R. E. Munn, Eds. (Cambridge University Press, Cambridge, 1986). chapt. 9.

7. R. T. Watson, et al, in Climate Change. The IPCC Scientific Assessment J. T. Houghton, G. J. Jenkins and J. J. Ephraums, Eds., (Cambridge University Press, 1990), chapt. 1.

8. P. M. Vitousek et al., *Science*, 277, 494, (1997).

9. E. O. Wilson, *The Diversity of Life*, Penguin Books, 1992.

regulatory measures to end their production had been taken. Coastal wetlands are also affected by humans, having resulted in the loss of 50% of the world's mangroves. Finally, mechanized human predation ("fisheries") removes more than 25% of the primary production of the oceans in the upwelling regions and 35% in the temperate continental shelf regions.[10] Anthropogenic effects are also well illustrated by the history of biotic communities that leave remains in lake sediments. The effects documented include modification of the geochemical cycle in large freshwater systems and occur in systems remote from primary sources.[11–13]

Considering these and many other major and still growing impacts of human activities on earth and atmosphere, and at all, including global, scales, it seems to us more than appropriate to emphasize the central role of mankind in geology and ecology by proposing to use the term "anthropocene" for the current geological epoch. The impacts of current human activities will continue over long periods. According to a study by Berger and Loutre,[14] because of the anthropogenic emissions of CO_2, climate may depart significantly from natural behavior over the next 50,000 years.

To assign a more specific date to the onset of the "anthropocene" seems somewhat arbitrary, but we propose the latter part of the 18th century, although we are aware that alternative proposals can be made (some may even want to include the entire holocene). However, we choose this date because, during the past two centuries, the global effects of human activities have become clearly noticeable. This is the period when data retrieved from glacial ice cores show the beginning of a growth in the atmospheric concentrations of several "greenhouse

---

10. D. Pauly and V. Christensen, *Nature*, 374, 255–57, 1995.

11. E. F. Stoermer and J. P. Smol, Eds., *The Diatoms: Applications for the Environmental and Earth Sciences* (Cambridge University Press, Cambridge, 1999).

12. C. L. Schelske and E. F. Stoermer, *Science*, 173 (1971); D. Verschuren et al., *J. Great Lakes Res.*, 24 (1998).

13. M. S. V. Douglas, J. P. Smol and W. Blake Jr., *Science* 266 (1994).

14. A. Berger and M.-F. Loutre, *C. R. Acad. Sci. Paris*, 323, II A, 1–16, 1996.

gases," in particular CO_2 and CH_4.[15] Such a starting date also coincides with James Watt's invention of the steam engine in 1784. About at that time, biotic assemblages in most lakes began to show large changes.[16]

Without major catastrophes like an enormous volcanic eruption, an unexpected epidemic, a large-scale nuclear war, an asteroid impact, a new ice age, or continued plundering of Earth's resources by partially still primitive technology (the last four dangers can, however, be prevented in a real functioning noösphere) mankind will remain a major geological force for many millennia, maybe millions of years, to come. To develop a worldwide accepted strategy leading to sustainability of ecosystems against human induced stresses will be one of the great future tasks of mankind, requiring intensive research efforts and wise application of the knowledge thus acquired in the noösphere, better known as knowledge or information society. An exciting, but also difficult and daunting task lies ahead of the global research and engineering community to guide mankind towards global, sustainable, environmental management.[17]

15. See note 7 above.
16. See notes 11–13 above.
17. J. Schellnhuber, *Nature*, 402, C19–C23, 1999.

MICHAEL SHELLENBERGER AND TED NORDHAUS

THE DEATH OF ENVIRONMENTALISM

Global Warming Politics in a Post-Environmental World

In 2003, as mainstream environmental organizations began to turn to the courts to try to force a recalcitrant President George W. Bush to regulate greenhouse gases, longtime environmental strategists Michael Shellenberger and Ted Nordhaus cofounded a new environmental think tank called the Breakthrough Institute, focused on pragmatic, innovation-oriented solutions to climate change and other environmental problems. Rather than focusing on regulation and measures to make energy, carbon emissions, and pollution more expensive, the Breakthrough Institute argued that national-level climate policy should endeavor to promote technology that would make clean energy cheaper. The organization's first publication, "The Death of Environmentalism: Global Warming in a Post-Environmental World," assessed the conceptual and institutional limitations of American environmentalism in dealing with the problem of climate change. What, for Nordhaus and Shellenberger, had climate change changed? Why might their approach have been particularly appealing to other environmentalists and a broader public concerned about climate change in 2003?

Michael Shellenberger and Ted Nordhaus, "The Death of Environmentalism: Global Warming Politics in a Post-Environmental World," Breakthrough Institute, 2004. Excerpt: pp. 6–7, 8, 10, 14–15, 25, 26, 28, 33, 34. www.thebreakthrough.org/images/Death_of_Environmentalism.pdf. By permission of the Breakthrough Institute.

Those of us who are children of the environmental movement must never forget that we are standing on the shoulders of all those who came before us.

The clean water we drink, the clean air we breathe, and the protected wilderness we treasure are all, in no small part, thanks to them. The two of us have worked for most of the country's leading environmental organizations as staff or consultants. We hold a sincere and abiding respect for our parents and elders in the environmental community. They have worked hard and accomplished a great deal. For that we are deeply grateful.

At the same time, we believe that the best way to honor their achievements is to acknowledge that modern environmentalism is no longer capable of dealing with the world's most serious ecological crisis.

Over the last 15 years environmental foundations and organizations have invested hundreds of millions of dollars into combating global warming.

We have strikingly little to show for it.

From the battles over higher fuel efficiency for cars and trucks to the attempts to reduce carbon emissions through international treaties, environmental groups repeatedly have tried and failed to win national legislation that would reduce the threat of global warming. As a result, people in the environmental movement today find themselves politically less powerful than we were one and a half decades ago.

Yet in lengthy conversations, the vast majority of leaders from the largest environmental organizations and foundations in the country insisted to us that we are on the right track.

Nearly all of the more than two-dozen environmentalists we interviewed underscored that climate change demands that we remake the global economy in ways that will transform the lives of six billion people. All recognize that it's an undertaking of monumental size and complexity. And all acknowledged that we must reduce emissions by up to 70 percent as soon as possible.

But in their public campaigns, not one of America's environmental leaders is articulating a vision of the future commensurate with the magnitude of the crisis. Instead they are promoting technical policy fixes like pollution controls and higher vehicle mileage standards—proposals that provide neither the popular inspiration nor the political alliances the community needs to deal with the problem.

By failing to question their most basic assumptions about the problem and the solution, environmental leaders are like generals fighting the last war—in particular the war they fought and won for basic environmental protections more than 30 years ago. It was then that the community's political strategy became defined around using science to define the problem as "environmental" and crafting technical policy proposals as solutions.

Our thesis is this: the environmental community's narrow definition of its self-interest leads to a kind of policy literalism that undermines its power. When you look at the long string of global warming defeats under Presidents Bill Clinton and George W. Bush, it is hard not to conclude that the environmental movement's approach to problems and policies hasn't worked particularly well. And yet there is nothing about the behavior of environmental groups, and nothing in our interviews with environmental leaders, that indicates that we as a community are ready to think differently about our work.

What the environmental movement needs more than anything else right now is to take a collective step back to rethink everything. We will never be able to turn things around as long as we understand our failures as essentially tactical, and make proposals that are essentially technical....

Those of us who were children during the birth of the modern environmental movement have no idea what it feels like to really win big.

Our parents and elders experienced something during the 1960s and 70s that today seems like a dream: the passage of a series of powerful environmental laws too numerous to list, from the Endangered Species Act to the Clean Air and Clean Water Acts to the National Environmental Policy Act.

Experiencing such epic victories had a searing impact on the minds of the movement's founders. It established a way of thinking about the environment and politics that has lasted until today.

It was also then, at the height of the movement's success, that the seeds of failure were planted. The environmental community's success created a strong confidence—and in some cases bald arrogance—that the environmental protection frame was enough to succeed at a policy level. The environmental community's belief that their power derives from defining themselves as defenders of "the environment" has prevented us from winning major legislation on global warming at the national level.

We believe that the environmental movement's foundational concepts, its method for framing legislative proposals, and *its very institutions* are outmoded. Today environmentalism is just another special interest. Evidence for this can be found in its concepts, its proposals, and its reasoning. What stands out is how arbitrary environmental leaders are about what gets counted and what doesn't as "environmental." Most of the movement's leading thinkers, funders and advocates do not question their most basic assumptions about who we are, what we stand for, and what it is that we should be doing. . . .

The entire landscape in which politics plays out has changed radically in the last 30 years, yet the environmental movement acts as though proposals based on "sound science" will be sufficient to overcome ideological and industry opposition. Environmentalists are in a culture war whether we like it or not. It's a war over our core values as Americans and over our vision for the future, and it won't be won by appealing to the rational consideration of our collective self-interest.

We have become convinced that modern environmentalism, with all of its unexamined assumptions, outdated concepts and exhausted strategies, must die so that something new can live. Those of us who pay so much attention to nature's cycles know better than to fear death, which is inseparable from life. In the words of the *Tao Ti Ching*, "If you aren't afraid of dying there is nothing you can't achieve." . . .

What do we worry about when we worry about global warming? Is it the refugee crisis that will be caused when Caribbean nations are flooded? If so, shouldn't our focus be on building bigger sea walls and disaster preparedness?

Is it the food shortages that will result from reduced agricultural production? If so, shouldn't our focus be on increasing food production?

Is it the potential collapse of the Gulf Stream, which could freeze upper North America and northern Europe and trigger, as a recent Pentagon scenario suggests, world war?

Most environmental leaders would scoff at such framings of the problem and retort, "Disaster preparedness is not an environmental problem." It is a hallmark of environmental rationality to believe that we environmentalists search for "root causes" not "symptoms." What, then, is the cause of global warming?

For most within the environmental community, the answer is easy: too much carbon in the atmosphere. Framed this way, the solution is logical: we need to pass legislation that reduces carbon emissions. But what are the obstacles to removing carbon from the atmosphere?

Consider what would happen if we identified the obstacles as:

- The radical right's control of all three branches of the US government.
- Trade policies that undermine environmental protections.
- Our failure to articulate an inspiring and positive vision.
- Overpopulation.
- The influence of money in American politics.
- Our inability to craft legislative proposals that shape the debate around core American values.
- Poverty.
- Old assumptions about what the problem is and what it isn't.

The point here is not just that global warming has many causes but also that the solutions we dream up depend on how we structure the problem.

The environmental movement's failure to craft inspiring and powerful proposals to deal with global warming is directly related to the movement's reductive logic about the supposedly root causes (e.g., "too much carbon in the atmosphere") of any given environmental problem. The problem is that once you identify something as the root cause, you have little reason to look for even deeper causes or connections with other root causes. . . .

The environmental movement's technical policy orientation has created a kind of myopia: everyone is looking for short-term policy pay-off.

We could find nobody who is crafting political proposals that, through the alternative vision and values they introduce, create the context for electoral and legislative victories down the road. Almost every environmental leader we interviewed is focused on short-term policy work, not long-term strategies....

PART II: GOING BEYOND SPECIAL INTERESTS AND SINGLE ISSUES

The marriage between vision, values, and policy has proved elusive for environmentalists. Most environmental leaders, even the most vision-oriented, are struggling to articulate proposals that have coherence. This is a crisis because environmentalism will never be able to muster the strength it needs to deal with the global warming problem as long as it is seen as a "special interest." And it will continue to be seen as a special interest as long as it narrowly identifies the problem as "environmental" and the solutions as technical....

Whereas neocons make proposals using their core values as a strategy for building a political majority, liberals, especially environmentalists, try to win on one issue at a time. We come together only around elections when our candidates run on our issue lists and technical policy solutions. The problem, of course, isn't just that environmentalism has become a special interest. The problem is that all liberal politics have become special interests....

If environmentalists hope to become more than a special interest we must start framing our proposals around core American values and start seeing our own values as central to what motivates and guides our politics. Doing so is crucial if we are to build the political momentum—a sustaining movement—to pass and implement the legislation that will achieve action on global warming and other issues....

Environmentalists need to tap into the creative worlds of myth-making, even religion, not to better sell narrow and technical policy proposals but rather to figure out who we are and who we need to be.

Above all else, we need to take a hard look at the institutions the movement has built over the last 30 years. Are existing environmental institutions up to the task of imagining the post-global warming world? Or do we now need a set of new institutions founded around a more expansive vision and set of values?

NICHOLAS STERN

STERN REVIEW ON THE ECONOMICS OF CLIMATE CHANGE

"The Death of Environmentalism" took a hard look at environmentalists' strategies in a time of presidential retrenchment on environmental issues. While the George W. Bush White House resisted even modest attempts to incorporate climate research into federal policy in the early 2000s, however, other Western governments continued their efforts to make sense of what climate change might mean for their people and their economies. In 2005, the head of the British Treasury, chancellor of the exchequer (and later prime minister) Gordon Brown, commissioned London School of Economics economist Sir Nicholas Stern to lead a team of treasury economists in creating a report on the effects of global warming on the world economy. Released in October of 2006, the seven-hundred-page report called climate change the greatest and widest-ranging market failure the world has ever seen. According to the report, the overall costs of climate change impacts will exceed 5 percent of GDP per year for the foreseeable future, reaching as high as 20 percent of GDP or more in some modeling scenarios. Here, Stern outlines the task of the group and the summary conclusions of the report. As a report commissioned by the British government, the Stern Review did not pass through peer review, and it met with mixed reviews from other economists. Perhaps most vocal among the report's critics was American economist William Nordhaus—the same William Nordhaus included in the 1983 Nierenberg Report for the National Academy of

---

Nicholas Stern, "Stern Review on the Economics of Climate Change," prepublication edition, HM Treasury, 2006. Excerpt: Introduction, pp. iv–v; Executive Summary, pp. i–ii, viii–ix, xi, vi, vii, ix.

Sciences excerpted in part 4. As Nordhaus explains in his review of the Stern Review (excerpted below), one of the sticking points for economists critical of Stern was the "discount rate," in this case the difference in value between money spent on climate change today and money spent on climate change in the future. How do both authors situate economics relative to climate science in their treatment of climate change economics here? How do they each deal with questions of uncertainty?

• •

INTRODUCTION

The economics of climate change is shaped by the science. That is what dictates the structure of the economic analysis and policies; therefore we start with the science.

Human-induced climate change is caused by the emissions of carbon dioxide and other greenhouse gases (GHGs) that have accumulated in the atmosphere mainly over the past 100 years.

The scientific evidence that climate change is a serious and urgent issue is now compelling. It warrants strong action to reduce greenhouse gas emissions around the world to reduce the risk of very damaging and potentially irreversible impacts on ecosystems, societies and economies. With good policies the costs of action need not be prohibitive and would be much smaller than the damage averted.

Reversing the trend to higher global temperatures requires an urgent, world-wide shift towards a low-carbon economy. Delay makes the problem much more difficult and action to deal with it much more costly. Managing that transition effectively and efficiently poses ethical and economic challenges, but also opportunities, which this Review sets out to explore.

Economics has much to say about assessing and managing the risks of climate change, and about how to design national and international responses for both the reduction of emissions and adaptation to the impacts that we can no longer avoid. If economics is used to design cost-effective policies, then taking action to tackle climate change will enable societies' potential for well-being to increase much faster in the long run than without action; we can be "green" and grow. Indeed, if we are not "green," we will eventually undermine growth, however measured.

This Review takes an international perspective on the economics of climate change. Climate change is a global issue that requires a global response. The science tells us that emissions have the same effects from wherever they arise. The implication for the economics is that this is clearly and unambiguously an international collective action problem with all the attendant difficulties of generating coherent action and of avoiding free riding. It is a problem requiring international cooperation and leadership.

Our approach emphasizes a number of key themes, which will feature throughout.

- We use a consistent approach towards *uncertainty*. The science of climate change is reliable, and the direction is clear. But we do not know precisely when and where particular impacts will occur. Uncertainty about impacts strengthens the argument for mitigation: this Review is about the economics of the management of very large risks.
- We focus on a quantitative understanding of *risk*, assisted by recent advances in the science that have begun to assign probabilities to the relationships between emissions and changes in the climate system, and to those between the climate and the natural environment.
- We take a systematic approach to the treatment of inter- and intra-generational *equity* in our analysis, informed by a consideration of what various ethical perspectives imply in the context of climate change. Inaction now risks great damage to the prospects of future generations, and particularly to the poorest amongst them. A coherent economic analysis of policy requires that we be explicit about the effects.

Economists describe human-induced climate change as an "externality" and the global climate as a "public good." Those who create greenhouse gas emissions as they generate electricity, power their factories, flare off gases, cut down forests, fly in planes, heat their homes or drive their cars do not have to pay for the costs of the climate change that results from their contribution to the accumulation of those gases in the atmosphere.

But climate change has a number of features that together distinguish it from other externalities. It is global in its causes and consequences; the impacts of climate change are persistent and develop over the long run; there are uncertainties that prevent precise quantification of the economic impacts; and there is a serious risk of major, irreversible change with non-marginal economic effects.

This analysis leads us to five sets of questions that shape Parts 2 to 6 of the Review.

- What is our understanding of the risks of the impacts of climate change, their costs, and on whom they fall?
- What are the options for reducing greenhouse-gas emissions, and what do they cost? What does this mean for the economics of the choice of paths to stabilization for the world? What are the economic opportunities generated by action on reducing emissions and adopting new technologies?
- For mitigation of climate change, what kind of incentive structures and policies will be most effective, efficient and equitable? What are the implications for the public finances?
- For adaptation, what approaches are appropriate and how should they be financed?
- How can approaches to both mitigation and adaptation work at an international level?

EXECUTIVE SUMMARY

... Climate change presents a unique challenge for economics: it is the greatest and widest-ranging market failure ever seen. The economic analysis must therefore be global, deal with long time horizons, have the economics of risk and uncertainty at centre stage, and examine the possibility of major, non-marginal change. To meet these requirements, the Review draws on ideas and techniques from most of the important areas of economics, including many recent advances.

The Benefits of Strong, Early Action on Climate Change Outweigh the Costs

The effects of our actions now on future changes in the climate have long lead times. What we do now can have only a limited effect on the climate over the next 40 or 50 years. On the other hand what we do in the next 10 or 20 years can have a profound effect on the climate in the second half of this century and in the next.

No-one can predict the consequences of climate change with complete certainty; but we now know enough to understand the risks. Mitigation—taking strong action to reduce emissions—must be viewed as an investment, a cost incurred now and in the coming few decades to avoid the risks of very severe consequences in the future. If these investments are made wisely, the costs will be manageable, and there will be a wide range of opportunities for growth and development along the way. For this to work well, policy must promote sound market signals, overcome market failures and have equity and risk mitigation at its core. That essentially is the conceptual framework of this Review.

The Review considers the economic costs of the impacts of climate change, and the costs and benefits of action to reduce the emissions of greenhouse gases (GHGs) that cause it, in three different ways:

- Using disaggregated techniques, in other words considering the physical impacts of climate change on the economy, on human life and on the environment, and examining the resource costs of different technologies and strategies to reduce greenhouse gas emissions;
- Using economic models, including integrated assessment models that estimate the economic impacts of climate change, and macro-economic models that represent the costs and effects of the transition to low-carbon energy systems for the economy as a whole;
- Using comparisons of the current level and future trajectories of the "social cost of carbon" (the cost of impacts associated with an additional unit of greenhouse gas emissions) with the marginal abatement cost (the costs associated with incremental reductions in units of emissions).

From all of these perspectives, the evidence gathered by the Review leads to a simple conclusion: the benefits of strong, early action considerably outweigh the costs.

The evidence shows that ignoring climate change will eventually damage economic growth. Our actions over the coming few decades could create risks of major disruption to economic and social activity, later in this century and in the next, on a scale similar to those associated with the great wars and the economic depression of the first half of the 20th century. And it will be difficult or impossible to reverse these changes. Tackling climate change is the pro-growth strategy for the longer term, and it can be done in a way that does not cap the aspirations for growth of rich or poor countries. The earlier effective action is taken, the less costly it will be. . . .

Most formal modeling in the past has used as a starting point a scenario of 2–3°C warming. In this temperature range, the cost of climate change could be equivalent to a permanent loss of around 0–3% in global world output compared with what could have been achieved in a world without climate change. Developing countries will suffer even higher costs.

However, those earlier models were too optimistic about warming: more recent evidence indicates that temperature changes resulting from BAU [business as usual] trends in emissions may exceed 2–3°C by the end of this century. This increases the likelihood of a wider range of impacts than previously considered. Many of these impacts, such as abrupt and large-scale climate change, are more difficult to quantify. With 5–6°C warming—which is a real possibility for the next century—existing models that include the risk of abrupt and large-scale climate change estimate an average 5–10% loss in global GDP, with poor countries suffering costs in excess of 10% of GDP. Further, there is some evidence of small but significant risks of temperature rises even above this range. Such temperature increases would take us into territory unknown to human experience and involve radical changes in the world around us. . . .

CO_2 emissions per head have been strongly correlated with GDP per head. As a result, since 1850, North America and Europe have produced around 70% of all the CO_2 emissions due to energy production, while developing countries have accounted for less than one quarter. Most future emissions growth will come from today's developing countries, because of their more rapid population and GDP growth and their increasing share of energy-intensive industries.

Yet despite the historical pattern and the BAU projections, the world does not need to choose between averting climate change and promoting growth and development. Changes in energy technologies and the structure of economies have reduced the responsiveness of emissions to income growth, particularly in some of the richest countries. With strong, deliberate policy choices, it is possible to "decarbonise" both developed and developing economies on the scale required for climate stabilization, while maintaining economic growth in both.

SUMMARY OF CONCLUSIONS

There is still time to avoid the worst impacts of climate change, if we take strong action now.

This Review has assessed a wide range of evidence on the impacts of climate change and on the economic costs, and has used a number of different techniques to assess costs and risks. From all of these perspectives, the evidence gathered by the Review leads to a simple conclusion: the benefits of strong and early action far outweigh the economic costs of not acting.

Using the results from formal economic models, the Review estimates that if we don't act, the overall costs and risks of climate change will be equivalent to losing at least 5% of global GDP each year, now and forever. If a wider range of risks and impacts is taken into account, the estimates of damage could rise to 20% of GDP or more.

In contrast, the costs of action—reducing greenhouse gas emissions to avoid the worst impacts of climate change—can be limited to around 1% of global GDP each year.

The investment that takes place in the next 10–20 years will have a profound effect on the climate in the second half of this century and in the next. Our actions now and over the coming decades could create risks of major disruption to economic and social activity, on a scale similar to those associated with the great wars and the economic depression of the first half of the 20th century. And it will be difficult or impossible to reverse these changes. . . .

The risks of the worst impacts of climate change can be substantially reduced if greenhouse gas levels in the atmosphere can be stabilized between 450 and 550ppm CO_2 equivalent (CO_2e). The current level is 430ppm CO_2e today, and it is rising at more than 2ppm each year. Stabilization in this range would require emissions to be at least 25% below current levels by 2050, and perhaps much more.

Ultimately, stabilization—at whatever level—requires that annual emissions be brought down to more than 80% below current levels. . . .

The costs of taking action are not evenly distributed across sectors or around the world. Even if the rich world takes on responsibility for absolute cuts in emissions of 60–80% by 2050, developing countries must take significant action too. But developing countries should not be required to bear the full costs of this action alone, and they will not have to. Carbon markets in rich countries are already beginning to deliver flows of finance to support low-carbon development, including through the Clean Development Mechanism. A transformation of these flows is now required to support action on the scale required.

Action on climate change will also create significant business opportunities, as new markets are created in low-carbon energy technologies and other low-carbon goods and services. These markets could grow to be worth hundreds of billions of dollars each year, and employment in these sectors will expand accordingly.

The world does not need to choose between averting climate change and promoting growth and development. Changes in energy technologies and in the structure of economies have created opportunities to decouple growth from greenhouse gas emissions. Indeed, ignoring climate change will eventually damage economic growth.

Tackling climate change is the pro-growth strategy for the longer term, and it can be done in a way that does not cap the aspirations for growth of rich or poor countries.

WILLIAM D. NORDHAUS

A REVIEW OF THE *STERN REVIEW* ON THE ECONOMICS OF CLIMATE CHANGE

How much and how fast should we react to the threat of global warming? The *Stern Review* argues that the damages from climate change are large, and that nations should undertake sharp and immediate reductions in greenhouse gas emissions. An examination of the *Review*'s radical revision of the economics of climate change finds, however, that it depends decisively on the assumption of a near-zero time discount rate combined with a specific utility function. The *Review*'s unambiguous conclusions about the need for extreme immediate action will not survive the substitution of assumptions that are consistent with today's marketplace real interest rates and savings rates.

1. OPPOSITE ENDS OF THE GLOBE

It appears that no two places on earth are further apart on global warming policies than the White House and 10 Downing Street. In 2001, President George W. Bush announced his opposition to binding constraints on greenhouse gas emissions. In his letter of opposition, he stated, "I oppose the Kyoto Protocol because it exempts 80 percent of the world, including major population centers such as China and India, from compliance, and would cause serious harm to the U.S. economy"

William D. Nordhaus, "A Review of the *Stern Review* on the Economics of Climate Change," *Journal of Economic Literature* 45 (September 2007): 686–87, 688, 689, 701.

(Bush 2001). This policy, much like the war in Iraq, was undertaken with no discernible economic analysis.[1]

In stark contrast, the British government in November 2006 presented a comprehensive new study, the *Stern Review on the Economics of Climate Change* (hereafter the *Review*).[2] Prime Minister Tony Blair painted a dark picture for the globe at its unveiling, "It is not in doubt that if the science is right, the consequences for our planet are literally disastrous.... [W]ithout radical international measures to reduce carbon emissions within the next 10 to 15 years, there is compelling evidence to suggest we might lose the chance to control temperature rises" (Blair 2006).

The summary in the *Review* was equally stark: "[T]he *Review* estimates that if we don't act, the overall costs and risks of climate change will be equivalent to losing at least 5% of global GDP each year, now and forever. If a wider range of risks and impacts is taken into account, the estimates of damage could rise to 20% of GDP or more.... Our actions now and over the coming decades could create risks... on a scale similar to those associated with the great wars and the economic depression of the first half of the 20th century" (p. xv).

These results are dramatically different from earlier economic models that use the same basic data and analytical structure. One of the major findings in the economics of climate change has been that efficient or "optimal" economic policies to slow climate change involve modest rates of emissions reductions in the near term, followed by sharp reductions in the medium and long term. We might call this the

---

1. There is no record of a fact sheet or other economic analysis accompanying the letter. The Bush Administration's economic analysis was contained in the *Economic Report of the President and the Council of Economic Advisers* (2002), chapter 6, published almost a year after President Bush's letter to the Senators. The *Economic Report*'s analysis suggests that the Kyoto Protocol is costly, but its analysis does not show that binding action is economically unwarranted.

2. The printed version is Nicholas Stern (2007). Also, see the electronic edition at that reference. It is assumed that the printed version is the report of record, and all citations are to the printed version. The printed version contains a "Postscript" which is in part a response to the early critics, including a response to the November 17, 2006, draft of this review.

climate-policy ramp, in which policies to slow global warming increasingly tighten or ramp up over time.[3] ...

The logic of the climate-policy ramp is straightforward. In a world where capital is productive, the highest-return investments today are primarily in tangible, technological, and human capital, including research and development on low-carbon technologies. In the coming decades, damages are predicted to rise relative to output. As that occurs, it becomes efficient to shift investments toward more intensive emissions reductions. The exact mix and timing of emissions reductions depend upon details of costs, damages, and the extent to which climate change and damages are nonlinear and irreversible.

There are many perils, costs, and uncertainties—known unknowns as well as unknown unknowns—involved in unchecked climate change.[4] Economic analyses have searched for strategies that will balance the costs of action with the perils of inaction. All economic studies find a case for imposing immediate restraints on greenhouse gas emissions, but the difficult questions are how much and how fast. The *Review* is in the tradition of economic cost-benefit analyses, but it has strikingly different conclusions from the mainstream economic models.[5] Because it has conclusions that are so different from most economic studies, the present note examines the reasons for this major difference. Is this radical revision of global-warming economics warranted? What are the reasons for the difference?[6] ...

---

3. This strategy is a hallmark of virtually every study of intertemporal efficiency in climate-change policy. It was one of the major conclusions in a review of integrated-assessment models: "Perhaps the most surprising result is the consensus that given calibrated interest rates and low future economic growth, modest controls are generally optimal" (David L. Kelly and Charles D. Kolstad 1999). A survey of the results of greenhouse-gas stabilization in several models in contained in Energy Modeling Forum Study 19 (2004). This result has been found in all five generations of the Yale/DICE/RICE global-warming models developed over the 1975–2007 period; see the references in footnote 28.

4. See James Hansen et al. (2006) for a recent warning.

5. An early precursor of this Review is the study by William R. Cline (1992). Cline's analysis of discounting was virtually identical to that in the *Review*.

6. There is by spring 2007 a large body of commentary on the *Stern Review*, including the companion article by Martin Weitzman in this issue. A critical

The scientific ground rules of government reports produced by professional scientists and economists are not codified. My vantage point, having been both producer and consumer of government reports, is that we expect them to be factually correct, present a professionally accurate representation of the technical scientific issues, support the government's policies, but not necessarily to be a textbook with a balanced view of all competing theories. By this definition of the ground rules, the *Review* fits well within the boundaries. For the most part, it accurately describes the basic economic questions involved in global warming. However, it tends to emphasize studies and findings that support its policy recommendations, while reports with opposing views of the dangers of global warming are ignored. Such are the rules of the game, but we should be alert in reading the *Review* that—even though it was published by a university press—it is not standard academic analysis. . . .

This deviation from the norm of modern science does not necessarily discredit the *Review*, but it does mean that fatal flaws in evidence and reasoning, which might have been caught in the early stages under normal ground rules, may emerge after the report has been published. . . .

But these points are not the nub of the matter. Rather, the *Review*'s radical view of policy stems from an extreme assumption about discounting. Discounting is a factor in climate-change policy—indeed in all investment decisions—that involves the relative weight of future and present payoffs. At first blush, this area would seem a technicality. Unfortunately, it cannot be buried in a footnote, for discounting is central to the radical revision. The *Review* proposes ethical assumptions

---

discussion of key assumptions is provided in Richard S. J. Tol and Gary W. Yohe (2006) and Robert O. Mendelsohn (2006). A particularly useful discussion of discounting issues is contained in Partha Dasgupta (2007). An analysis which focuses on the extreme findings of the *Review* is S. Niggol Seo (2006). A discussion of ethics is in Wilfred Beckerman and Cameron Hepburn (2007). A sensitivity analysis of the ethical parameters with much the same message as the present article is Sergey Mityakov (2007). A wide-ranging attack on various elements is contained in Robert M. Carter et al. (2006) and Ian Byatt et al. (2006). Insurance issues and discounting are discussed in Christian Gollier (2006).

that produce very low discount rates. Combined with other assumptions, this magnifies impacts in the distant future and rationalizes deep cuts in emissions, and indeed in all consumption, today. If we substitute more conventional discount rates used in other global-warming analyses, by governments, by consumers, or by businesses, the *Review*'s dramatic results disappear, and we come back to the climate-policy ramp described above. The balance of this discussion focuses on this central issue....

5. SUMMARY VERDICT

How much and how fast should the globe reduce greenhouse gas emissions? How should nations balance the costs of the reductions against the damages and dangers of climate change? The *Stern Review* answers these questions clearly and unambiguously: we need urgent, sharp, and immediate reductions in greenhouse gas emissions. I am reminded here of President Harry Truman's complaint that his economists would always say, on the one hand this and on the other hand that. He wanted a one-handed economist. The *Stern Review* is a President's or a Prime Minister's dream come true. It provides decisive answers instead of the dreaded conjectures, contingencies, and qualifications. However, a closer look reveals that there is indeed another hand to these answers. The *Review*'s radical revision of the economics of climate change does not arise from any new economics, science, or modeling. Rather, it depends decisively on the assumption of a near-zero time discount rate combined with a specific utility function. The *Review*'s unambiguous conclusions about the need for extreme immediate action will not survive the substitution of assumptions that are more consistent with today's marketplace real interest rates and savings rates. Hence, the central questions about global-warming policy—how much, how fast, and how costly—remain open. The *Review* informs but does not answer these fundamental questions.

MASSACHUSETTS V. ENVIRONMENTAL PROTECTION AGENCY

In 2007, the Supreme Court decided in favor of the State of Massachusetts, alongside eleven other states, several U.S. cities, and a number of private organizations in their bid to petition the U.S. Environmental Protection Agency to regulate CO_2 and other greenhouse gases as pollutants. The case arose in 2003, when Massachusetts petitioned the EPA to regulate CO_2 from vehicle emissions under the 1990 Clean Air Act. Perhaps not surprisingly, the context of the petition is key to understanding the case. Recognizing the George W. Bush administration's recalcitrance on the Kyoto Protocol and other climate change policy, Massachusetts and its copetitioners turned to the courts to try and capitalize on existing legislation to force the EPA's hand on fuel efficiency standards. Over the next four years, the case moved through the courts, focusing on three issues. First, the petitioners had to demonstrate that they had "standing," or the right to claim injury from the nonregulation of greenhouse gases. Second, once the court granted Massachusetts standing based largely on its vulnerability to climate-induced sea level rise, the court had to decide whether CO_2 qualified as a "pollutant" under the Clean Air Act. And third, if CO_2 did qualify as a pollutant, the court had to determine whether scientific uncertainty gave the EPA the discretionary authority not to regulate CO_2. Ultimately, the court decided in favor of Massachusetts on all three issues, but Justices Roberts and Scalia authored strong dissenting opinions condemning the decision. How does the issue of climate change challenge both the majority and the minority in their legal interpretations of the main legal

---

Massachusetts v. Environmental Protection Agency, 549 U.S. 497 (2007). Excerpt: Majority Opinion, pp. 1–5, 6–7, 8, 9–10, 18, 20–21, 22–23, 25–26, 30–32; Dissent, pp. 1–2, 10, 11.

issues in the case? How does the court respond to questions of uncertainty and speculation here?

••

SUPREME COURT OF THE UNITED STATES

No. 05–1120

MASSACHUSETTS, ET AL., PETITIONERS *v.*
ENVIRONMENTAL PROTECTION AGENCY ET AL.

ON WRIT OF CERTIORARI TO THE UNITED STATES COURT
OF APPEALS FOR THE DISTRICT OF COLUMBIA CIRCUIT

[April 2, 2007]

JUSTICE STEVENS delivered the opinion of the Court.

A well-documented rise in global temperatures has coincided with a significant increase in the concentration of carbon dioxide in the atmosphere. Respected scientists believe the two trends are related. For when carbon dioxide is released into the atmosphere, it acts like the ceiling of a greenhouse, trapping solar energy and retarding the escape of reflected heat. It is therefore a species—the most important species—of a "greenhouse gas."

Calling global warming "the most pressing environmental challenge of our time,"[1] a group of States,[2] local governments,[3] and private organizations,[4] alleged in a petition for certiorari that the Environmental Protection Agency (EPA) has abdicated its responsibility under the

[1] Pet. for Cert. 22.
[2] California, Connecticut, Illinois, Maine, Massachusetts, New Jersey, New Mexico, New York, Oregon, Rhode Island, Vermont, and Washington.
[3] District of Columbia, American Samoa, New York City, and Baltimore.
[4] Center for Biological Diversity, Center for Food Safety, Conservation Law Foundation, Environmental Advocates, Environmental Defense, Friends of the Earth, Greenpeace, International Center for Technology Assessment, National Environmental Trust, Natural Resources Defense Council, Sierra Club, Union of Concerned Scientists, and U.S. Public Interest Research Group.

Clean Air Act to regulate the emissions of four greenhouse gases, including carbon dioxide. Specifically, petitioners asked us to answer two questions concerning the meaning of §202(a)(1) of the Act: whether EPA has the statutory authority to regulate greenhouse gas emissions from new motor vehicles; and if so, whether its stated reasons for refusing to do so are consistent with the statute.

In response, EPA, supported by 10 intervening States[5] and six trade associations,[6] correctly argued that we may not address those two questions unless at least one petitioner has standing to invoke our jurisdiction under Article III of the Constitution. Notwithstanding the serious character of that jurisdictional argument and the absence of any conflicting decisions construing §202(a)(1), the unusual importance of the underlying issue persuaded us to grant the writ. 548 U. S. __ (2006).

I

Section 202(a)(1) of the Clean Air Act, as added by Pub. L. 89-272, §101(8), 79 Stat. 992, and as amended by, inter alia, 84 Stat. 1690 and 91 Stat. 791, 42 U.S.C. §7521(a)(1), provides:

"The [EPA] Administrator shall by regulation prescribe (and from time to time revise) in accordance with the provisions of this section, standards applicable to the emission of any air pollutant from any class or classes of new motor vehicles or new motor vehicle engines, which in his judgment cause, or contribute to, air pollution which may reasonably be anticipated to endanger public health or welfare...."[7]

---

[5]Alaska, Idaho, Kansas, Michigan, Nebraska, North Dakota, Ohio, South Dakota, Texas, and Utah.

[6]Alliance of Automobile Manufacturers, National Automobile Dealers Association, Engine Manufacturers Association, Truck Manufacturers Association, CO_2 Litigation Group, and Utility Air Regulatory Group.

[7]The 1970 version of §202(a)(1) used the phrase "which endangers the public health or welfare" rather than the more protective "which may reasonably be anticipated to endanger public health or welfare." See §6(a) of the Clean Air Amendments of 1970, 84 Stat. 1690. Congress amended §202(a)(1) in 1977 to give its approval to the decision in *Ethyl Corp. v. EPA*, 541 F. 2d 1, 25 (CADC 1976) (en banc), which held that the Clean Air Act "and common sense . . . demand regulatory

The Act defines "air pollutant" to include "any air pollution agent or combination of such agents, including any physical, chemical, biological, radioactive . . . substance or matter which is emitted into or otherwise enters the ambient air." §7602(g). "Welfare" is also defined broadly: among other things, it includes "effects on . . . weather . . . and climate." §7602(h).

When Congress enacted these provisions, the study of climate change was in its infancy.[8] In 1959, shortly after the U.S. Weather Bureau began monitoring atmospheric carbon dioxide levels, an observatory in Mauna Loa, Hawaii, recorded a mean level of 316 parts per million. This was well above the highest carbon dioxide concentration—no more than 300 parts per million—revealed in the 420,000-year-old ice-core record.[9] By the time Congress drafted §202(a)(1) in 1970, carbon dioxide levels had reached 325 parts per million.[10]

In the late 1970s, the Federal Government began devoting serious

---

action to prevent harm, even if the regulator is less than certain that harm is otherwise inevitable." See §401(d)(1) of the Clean Air Act Amendments of 1977, 91 Stat. 791; see also H.R. Rep. No. 95-294, p. 49 (1977).

[8] The Council on Environmental Quality had issued a report in 1970 concluding that "[m]an may be changing his weather." Environmental Quality: The First Annual Report 93. Considerable uncertainty remained in those early years, and the issue went largely unmentioned in the congressional debate over the enactment of the Clean Air Act. But see 116 Cong. Rec. 32914 (1970) (statement of Sen. Boggs referring to Council's conclusion that "[a]ir pollution alters the climate and may produce global changes in temperature").

[9] See Intergovernmental Panel on Climate Change, Climate Change 2001: Synthesis Report, pp. 202–203 (2001). By drilling through thick Antarctic ice sheets and extracting "cores," scientists can examine ice from long ago and extract small samples of ancient air. That air can then be analyzed, yielding estimates of carbon dioxide levels. Ibid.

[10] A more dramatic rise was yet to come: In 2006, carbon dioxide levels reached 382 parts per million, see Dept. of Commerce, National Oceanic & Atmospheric Administration, Mauna Loa CO2 Monthly Mean Data, www.esrl.noaa.gov/gmd /ccgg/trends/co2_mm_mlo.dat (all Internet materials as visited Mar. 29, 2007, and available in Clerk of Court's case file), a level thought to exceed the concentration of carbon dioxide in the atmosphere at any point over the past 20-million years. See Intergovernmental Panel on Climate Change, Technical Summary of Working Group I Report 39 (2001).

attention to the possibility that carbon dioxide emissions associated with human activity could provoke climate change. In 1978, Congress enacted the National Climate Program Act, 92 Stat. 601, which required the President to establish a program to "assist the Nation and the world to understand and respond to natural and man-induced climate processes and their implications," id., §3. President Carter, in turn, asked the National Research Council, the working arm of the National Academy of Sciences, to investigate the subject. The Council's response was unequivocal: "If carbon dioxide continues to increase, the study group finds no reason to doubt that climate changes will result and no reason to believe that these changes will be negligible.... A wait-and-see policy may mean waiting until it is too late."[11]

Congress next addressed the issue in 1987, when it enacted the Global Climate Protection Act, Title XI of Pub. L. 100-204, 101 Stat. 1407, note following 15 U.S.C. §2901. Finding that "manmade pollution—the release of carbon dioxide, chlorofluorocarbons, methane, and other trace gases into the atmosphere—may be producing a long-term and substantial increase in the average temperature on Earth," §1102(1), 101 Stat. 1408, Congress directed EPA to propose to Congress a "coordinated national policy on global climate change," §1103(b), and ordered the Secretary of State to work "through the channels of multilateral diplomacy" and coordinate diplomatic efforts to combat global warming, §1103(c). Congress emphasized that "ongoing pollution and deforestation may be contributing now to an irreversible process" and that "[n]ecessary actions must be identified and implemented in time to protect the climate." §1102(4)....

On October 20, 1999, a group of 19 private organizations[15] filed a rulemaking petition asking EPA to regulate "greenhouse gas emissions from

---

[11]Climate Research Board, Carbon Dioxide and Climate: A Scientific Assessment, p. vii (1979).

[15]Alliance for Sustainable Communities; Applied Power Technologies, Inc.; Bio Fuels America; The California Solar Energy Industries Assn.; Clements Environmental Corp.; Environmental Advocates; Environmental and Energy Study Institute; Friends of the Earth; Full Circle Energy Project, Inc.; The Green Party

new motor vehicles under §202 of the Clean Air Act." App. 5. Petitioners maintained that 1998 was the "warmest year on record"; that carbon dioxide, methane, nitrous oxide, and hydrofluorocarbons are "heat trapping greenhouse gases"; that greenhouse gas emissions have significantly accelerated climate change; and that the IPCC's 1995 report warned that "carbon dioxide remains the most important contributor to [man-made] forcing of climate change." Id., at 13 (internal quotation marks omitted). The petition further alleged that climate change will have serious adverse effects on human health and the environment. Id., at 22–35. As to EPA's statutory authority, the petition observed that the agency itself had already confirmed that it had the power to regulate carbon dioxide. See id., at 18, n. 21. In 1998, Jonathan Z. Cannon, then EPA's General Counsel, prepared a legal opinion concluding that "CO_2 emissions are within the scope of EPA's authority to regulate," even as he recognized that EPA had so far declined to exercise that authority. . . .

On September 8, 2003, EPA entered an order denying the rulemaking petition. 68 Fed. Reg. 52922. The agency gave two reasons for its decision: (1) that contrary to the opinions of its former general counsels, the Clean Air Act does not authorize EPA to issue mandatory regulations to address global climate change, see id., at 52925–52929; and (2) that even if the agency had the authority to set greenhouse gas emission standards, it would be unwise to do so at this time, id., at 52929–52931.

In concluding that it lacked statutory authority over greenhouse gases, EPA observed that Congress "was well aware of the global climate change issue when it last comprehensively amended the [Clean Air Act] in 1990," yet it declined to adopt a proposed amendment establishing binding emissions limitations. Id., at 52926. . . . EPA further reasoned that Congress' "specially tailored solutions to global atmospheric issues," 68 Fed. Reg. 52926—in particular, its 1990 enactment

---

of Rhode Island; Greenpeace USA; International Center for Technology Assessment; Network for Environmental and Economic Responsibility of the United Church of Christ; New Jersey Environmental Watch; New Mexico Solar Energy Assn.; Oregon Environmental Council; Public Citizen; Solar Energy Industries Assn.; The SUN DAY Campaign. See App. 7–11.

of a comprehensive scheme to regulate pollutants that depleted the ozone layer, see Title VI, 104 Stat. 2649, 42 U. S. C. §§7671–7671q— counseled against reading the general authorization of §202(a)(1) to confer regulatory authority over greenhouse gases. . . .

EPA reasoned that climate change had its own "political history": Congress designed the original Clean Air Act to address local air pollutants rather than a substance that "is fairly consistent in its concentration throughout the *world's* atmosphere," 68 Fed. Reg. 52927 (emphasis added); declined in 1990 to enact proposed amendments to force EPA to set carbon dioxide emission standards for motor vehicles, *ibid.* (citing H.R. 5966, 101st Cong., 2d Sess. (1990); and addressed global climate change in other legislation, 68 Fed. Reg. 52927. . . . In essence, EPA concluded that climate change was so important that unless Congress spoke with exacting specificity, it could not have meant the agency to address it. . . .

Even assuming that it had authority over greenhouse gases, EPA explained in detail why it would refuse to exercise that authority. The agency began by recognizing that the concentration of greenhouse gases has dramatically increased as a result of human activities, and acknowledged the attendant increase in global surface air temperatures. *Id.*, at 52930. EPA nevertheless gave controlling importance to the NRC Report's statement that a causal link between the two "'cannot be unequivocally established.'" *Ibid.* (quoting NRC Report 17). Given that residual uncertainty, EPA concluded that regulating greenhouse gas emissions would be unwise. 68 Fed. Reg. 52930.

The Injury

The harms associated with climate change are serious and well recognized. Indeed, the NRC Report itself—which EPA regards as an "objective and independent assessment of the relevant science," 68 Fed. Reg. 52930—identifies a number of environmental changes that have already inflicted significant harms, including "the global retreat of mountain glaciers, reduction in snow-cover extent, the earlier spring melting of rivers and lakes, [and] the accelerated rate of rise of sea levels during the 20th century relative to the past few thousand years. . . . "

Causation

EPA does not dispute the existence of a causal connection between man-made greenhouse gas emissions and global warming. At a minimum, therefore, EPA's refusal to regulate such emissions "contributes" to Massachusetts' injuries.

EPA nevertheless maintains that its decision not to regulate greenhouse gas emissions from new motor vehicles contributes so insignificantly to petitioners' injuries that the agency cannot be hauled into federal court to answer for them. For the same reason, EPA does not believe that any realistic possibility exists that the relief petitioners seek would mitigate global climate change and remedy their injuries. That is especially so because predicted increases in greenhouse gas emissions from developing nations, particularly China and India, are likely to offset any marginal domestic decrease.

But EPA overstates its case. Its argument rests on the erroneous assumption that a small incremental step, because it is incremental, can never be attacked in a federal judicial forum. Yet accepting that premise would doom most challenges to regulatory action. Agencies, like legislatures, do not generally resolve massive problems in one fell regulatory swoop. See *Williamson v. Lee Optical of Okla., Inc.*, 348 U.S. 483, 489 (1955) ("[A] reform may take one step at a time, addressing itself to the phase of the problem which seems most acute to the legislative mind"). They instead whittle away at them over time, refining their preferred approach as circumstances change and as they develop a more nuanced understanding of how best to proceed. Cf. *SEC v. Chenery Corp.*, 332 U.S. 194, 202 (1947) ("Some principles must await their own development, while others must be adjusted to meet particular, unforeseeable situations"). That a first step might be tentative does not by itself support the notion that federal courts lack jurisdiction to determine whether that step conforms to law. . . .

The Remedy

While it may be true that regulating motor-vehicle emissions will not by itself *reverse* global warming, it by no means follows that we lack jurisdiction to decide whether EPA has a duty to take steps to *slow* or *reduce* it. See also *Larson v. Valente*, 456 U.S. 228, 244, n. 15 (1982) ("[A] plaintiff satisfies the redressability requirement when he shows that a favorable decision will relieve a discrete injury to himself. He need not show that a favorable decision will relieve his *every* injury"). Because of the enormity of the potential consequences associated with man-made climate change, the fact that the effectiveness of a remedy might be delayed during the (relatively short) time it takes for a new motor-vehicle fleet to replace an older one is essentially irrelevant.[23] Nor is it dispositive that developing countries such as China and India are poised to increase greenhouse gas emissions substantially over the next century: A reduction in domestic emissions would slow the pace of global emissions increases, no matter what happens elsewhere....

In sum—at least according to petitioners' uncontested affidavits—the rise in sea levels associated with global warming has already harmed and will continue to harm Massachusetts. The risk of catastrophic harm, though remote, is nevertheless real. That risk would be reduced to some extent if petitioners received the relief they seek. We therefore hold that petitioners have standing to challenge the EPA's denial of their rulemaking petition.[24]...

---

[23]See also *Mountain States Legal Foundation v. Glickman*, 92 F. 3d 1228, 1234 (CADC 1996) ("The more drastic the injury that government action makes more likely, the lesser the increment in probability to establish standing"); *Village of Elk Grove Village v. Evans*, 997 F. 2d 328, 329 (CA7 1993) ("[E]ven a small probability of injury is sufficient to create a case or controversy—to take a suit out of the category of the hypothetical—provided of course that the relief sought would, if granted, reduce the probability").

[24]In his dissent, THE CHIEF JUSTICE expresses disagreement with the Court's holding in *United States v. Students Challenging Regulatory Agency Procedures (SCRAP)*, 412 U.S. 669, 687–688 (1973). He does not, however, disavow this portion of Justice Stewart's opinion for the Court: "Unlike the specific and geographically limited federal action of which the petitioner complained in *Sierra Club* [v. *Morton*, 405 U. S. 727 (1972)], the challenged agency action in this case is applicable to substantially all of the Nation's railroads, and thus allegedly has an adverse environmental impact on all the natural resources of the country. Rather than a limited group of persons who used a picturesque valley in California, all persons who utilize the scenic resources of the country, and indeed all who breathe its air, could claim harm similar to that alleged by the environmental groups here. But we have already made it clear that standing is not to be denied

VI

On the merits, the first question is whether §202(a)(1) of the Clean Air Act authorizes EPA to regulate greenhouse gas emissions from new motor vehicles in the event that it forms a "judgment" that such emissions contribute to climate change. We have little trouble concluding that it does. In relevant part, §202(a)(1) provides that EPA "shall by regulation prescribe . . . standards applicable to the emission of any air pollutant from any class or classes of new motor vehicles or new motor vehicle engines, which in [the Administrator's] judgment cause, or contribute to, air pollution which may reasonably be anticipated to endanger public health or welfare." 42 U.S.C. §7521(a)(1).

Because EPA believes that Congress did not intend it to regulate substances that contribute to climate change, the agency maintains that carbon dioxide is not an "air pollutant" within the meaning of the provision.

The statutory text forecloses EPA's reading. The Clean Air Act's sweeping definition of "air pollutant" includes "*any* air pollution agent or combination of such agents, including any physical, chemical . . . substance or matter which is emitted into or otherwise enters the ambient air . . . ." §7602(g) (emphasis added). On its face, the definition embraces all airborne compounds of whatever stripe, and underscores that intent through the repeated use of the word "any."[25] Carbon dioxide, methane, nitrous oxide, and hydrofluorocarbons are without a doubt "physical [and] chemical . . . substance[s] which [are] emitted into . . . the ambient air." The statute is unambiguous.[26] . . .

---

simply because many people suffer the same injury. Indeed some of the cases on which we relied in *Sierra Club* demonstrated the patent fact that persons across the Nation could be adversely affected by major governmental actions. *To deny standing to persons who are in fact injured simply because many others are also injured, would mean that the most injurious and widespread Government actions could be questioned by nobody.* We cannot accept that conclusion." Ibid. (citations omitted and emphasis added).

It is moreover quite wrong to analogize the legal claim advanced by Massachusetts and the other public and private entities who challenge EPA's parsimonious construction of the Clean Air Act to a mere "lawyer's game." See post, at 14.

[25]See *Department of Housing and Urban Development v. Rucker*, 535 U.S. 125, 131 (2002) (observing that "'any' . . . has an expansive meaning, that is, one or some indiscriminately of whatever kind" (some internal quotation marks omitted)).

VII

The alternative basis for EPA's decision—that even if it does have statutory authority to regulate greenhouse gases, it would be unwise to do so at this time—rests on reasoning divorced from the statutory text.... Put another way, the use of the word "judgment" is not a roving license to ignore the statutory text. It is but a direction to exercise discretion within defined statutory limits.

If EPA makes a finding of endangerment, the Clean Air Act requires the agency to regulate emissions of the deleterious pollutant from new motor vehicles.... EPA no doubt has significant latitude as to the manner, timing, content, and coordination of its regulations with those of other agencies. But once EPA has responded to a petition for rulemaking, its reasons for action or inaction must conform to the authorizing statute. Under the clear terms of the Clean Air Act, EPA can avoid taking further action only if it determines that greenhouse gases do not contribute to climate change or if it provides some reasonable explanation as to why it cannot or will not exercise its discretion to determine whether they do....

EPA has refused to comply with this clear statutory command....

Nor can EPA avoid its statutory obligation by noting the uncertainty surrounding various features of climate change and concluding that it

[26]In dissent, JUSTICE SCALIA maintains that because greenhouse gases permeate the world's atmosphere rather than a limited area near the earth's surface, EPA's exclusion of greenhouse gases from the category of air pollution "agent[s]" is entitled to deference under *Chevron U.S.A. Inc. v. Natural Resources Defense Council, Inc.* 467 U.S. 837 (1984). See post, at 11–13. EPA's distinction, however, finds no support in the text of the statute, which uses the phrase "the ambient air" without distinguishing between atmospheric layers. Moreover, it is a plainly unreasonable reading of a sweeping statutory provision designed to capture "any physical, chemical ... substance or matter which is emitted into or otherwise enters the ambient air." 42 U.S.C. §7602(g). JUSTICE SCALIA does not (and cannot) explain why Congress would define "air pollutant" so carefully and so broadly, yet confer on EPA the authority to narrow that definition whenever expedient by asserting that a particular substance is not an "agent." At any rate, no party to this dispute contests that greenhouse gases both "ente[r] the ambient air" and tend to warm the atmosphere. They are therefore unquestionably "agent[s]" of air pollution.

would therefore be better not to regulate at this time. See 68 Fed. Reg. 52930–52931. If the scientific uncertainty is so profound that it precludes EPA from making a reasoned judgment as to whether greenhouse gases contribute to global warming, EPA must say so. That EPA would prefer not to regulate greenhouse gases because of some residual uncertainty— which, contrary to JUSTICE SCALIA's apparent belief, *post*, at 5–8, is in fact all that it said, see 68 Fed. Reg. 52929 ("We do not believe . . . that it would be either effective or appropriate for EPA *to establish [greenhouse gas] standards for motor vehicles at this time*" (emphasis added))—is irrelevant. The statutory question is whether sufficient information exists to make an endangerment finding.

In short, EPA has offered no reasoned explanation for its refusal to decide whether greenhouse gases cause or contribute to climate change. Its action was therefore "arbitrary, capricious, . . . or otherwise not in accordance with law." 42 U.S.C. §7607(d)(9)(A).

VIII

The judgment of the Court of Appeals is reversed, and the case is remanded for further proceedings consistent with this opinion.

It is so ordered.

CHIEF JUSTICE ROBERTS, with whom JUSTICE SCALIA, JUSTICE THOMAS, and JUSTICE ALITO join, dissenting.

Global warming may be a "crisis," even "the most pressing environmental problem of our time." Pet. for Cert. 26, 22. Indeed, it may ultimately affect nearly everyone on the planet in some potentially adverse way, and it may be that governments have done too little to address it. It is not a problem, however, that has escaped the attention of policymakers in the Executive and Legislative Branches of our Government, who continue to consider regulatory, legislative, and treaty-based means of addressing global climate change.

Apparently dissatisfied with the pace of progress on this issue in the elected branches, petitioners have come to the courts claiming broad-ranging injury, and attempting to tie that injury to the Government's alleged failure to comply with a rather narrow statutory provision. I would reject these challenges as nonjusticiable. Such a conclusion involves no judgment on whether global warming exists, what causes it, or the extent of the problem. Nor does it render petitioners without recourse. This Court's standing jurisprudence simply recognizes that redress of grievances of the sort at issue here "is the function of Congress and the Chief Executive," not the federal courts. *Lujan v. Defenders of Wildlife*, 504 U.S. 555, 576 (1992). I would vacate the judgment below and remand for dismissal of the petitions for review. . . .

The Court ignores the complexities of global warming, and does so by now disregarding the "particularized" injury it relied on in step one, and using the dire nature of global warming itself as a bootstrap for finding causation and redressability. First, it is important to recognize the extent of the emissions at issue here. Because local greenhouse gas emissions disperse throughout the atmosphere and remain there for anywhere from 50 to 200 years, it is global emissions data that are relevant. See App. to Pet. for Cert. A-73. According to one of petitioners' declarations, domestic motor vehicles contribute about 6 percent of global carbon dioxide emissions and 4 percent of global greenhouse gas emissions. Stdg. App. 232. The amount of global emissions at issue here is smaller still; §202(a)(1) of the Clean Air Act covers only *new* motor vehicles and *new* motor vehicle engines, so petitioners' desired emission standards might reduce only a fraction of 4 percent of global emissions. . . .

Petitioners are never able to trace their alleged injuries back through this complex web to the fractional amount of global emissions that might have been limited with EPA standards. In light of the bit-part domestic new motor vehicle greenhouse gas emissions have played in what petitioners describe as a 150-year global phenomenon, and the myriad additional factors bearing on petitioners' alleged injury—the loss of Massachusetts coastal land—the connection is far too speculative to establish causation. . . .

Petitioners' difficulty in demonstrating causation and redressability is not surprising given the evident mismatch between the source of their alleged injury— catastrophic global warming—and the narrow subject matter of the Clean Air Act provision at issue in this suit. The mismatch suggests that petitioners' true goal for this litigation may be more symbolic than anything else. The constitutional role of the courts, however, is to decide concrete cases—not to serve as a convenient forum for policy debates.

POPE FRANCIS

LAUDATO SI'

On Care for Our Common Home

Since the 1970s, the Catholic Church has periodically weighed in on the Earth's most pressing environmental problems. In 2015, however, Pope Francis became the first pope to dedicate an entire encyclical—a document sent out by the Holy See addressing Catholic doctrine for all those in communion—to the environment. *Laudato Si' (Praise Be to You): On the Care of Our Common Home,* is devoted to ecology and ecologically related themes and focuses heavily on the problem of climate change. Himself named after Saint Francis of Assisi, an Italian friar praised for his concern for nature and its nonhuman inhabitants, Pope Francis addressed the encyclical not primarily to the bishops of the Church, but to the entire planet. He also uncharacteristically framed the document as a dialogue, beginning with a significant review of some of the scientific components of global change. And yet, many of the components of a more traditional encyclical—the spiritual and humanitarian message, the appeal to both gospel and doctrine, the evangelical tone, and the prayers—remain, creating a nexus of Catholic teaching and environmental sensibility that has appealed to both Catholics and non-Catholics the world over.

· ·

Pope Francis, *Laudato Si': Encyclical Letter: On Care for Our Common Home,* sec. 2016. Excerpt: pp. 3, 4–6, 7, 8, 9–10, 11, 12, 13–14, 18–19, 32–33, 34–35, 44, 45–46, 48, 88, 118, 177–80. Available at http://w2.vatican.va/content/francesco/en/encyclicals/documents/papa-francesco_20150524_enciclica-laudato-si.html.

1. "Laudato si', mi' Signore"—"Praise be to you, my Lord." In the words of this beautiful canticle, Saint Francis of Assisi reminds us that our common home is like a sister with whom we share our life and a beautiful mother who opens her arms to embrace us. "Praise be to you, my Lord, through our Sister, Mother Earth, who sustains and governs us, and who produces various fruit with coloured flowers and herbs"[1] . . .

3. More than fifty years ago, with the world teetering on the brink of nuclear crisis, Pope Saint John XXIII wrote an Encyclical which not only rejected war but offered a proposal for peace. He addressed his message *Pacem in Terris* to the entire "Catholic world" and indeed "to all men and women of goodwill." Now, faced as we are with global environmental deterioration, I wish to address every person living on this planet. In my Apostolic Exhortation *Evangelii Gaudium*, I wrote to all the members of the Church with the aim of encouraging ongoing missionary renewal. In this Encyclical, I would like to enter into dialogue with all people about our common home.

4. In 1971, eight years after *Pacem in Terris*, Blessed Pope Paul VI referred to the ecological concern as "a tragic consequence" of unchecked human activity: "Due to an ill-considered exploitation of nature, humanity runs the risk of destroying it and becoming in turn a victim of this degradation."[2] He spoke in similar terms to the Food and Agriculture Organization of the United Nations about the potential for an "ecological catastrophe under the effective explosion of industrial civilization," and stressed "the urgent need for a radical change in the conduct of humanity," inasmuch as "the most extraordinary scientific advances, the most amazing technical abilities, the most astonishing economic growth, unless they are accompanied by authentic social and moral progress, will definitively turn against man."[3]

---

1. *Canticle of the Creatures*, in *Francis of Assisi: Early Documents*, vol. 1, New York-London-Manila, 1999, 113–114.
2. Apostolic Letter *Octogesima Adveniens* (14 May 1971), 21: *Acta Apostolicae Sedis* (AAS) 63 (1971), 416–417.
3. *Address to FAO on the 25th Anniversary of its Institution* (16 November 1970), 4: *AAS* 62 (1970), 833.

5. Saint John Paul II became increasingly concerned about this issue. In his first Encyclical he warned that human beings frequently seem "to see no other meaning in their natural environment than what serves for immediate use and consumption."[4] Subsequently, he would call for a global ecological *conversion*.[5] At the same time, he noted that little effort had been made to "safe-guard the moral conditions for an authentic *human ecology*."[6] The destruction of the human environment is extremely serious, not only because God has entrusted the world to us men and women, but because human life is itself a gift which must be defended from various forms of debasement. Every effort to protect and improve our world entails profound changes in "lifestyles, models of production and consumption, and the established structures of power which today govern societies."[7] Authentic human development has a moral character. It presumes full respect for the human person, but it must also be concerned for the world around us and "take into account the nature of each being and of its mutual connection in an ordered system."[8] Accordingly, our human ability to transform reality must proceed in line with God's original gift of all that is.[9]

6. My predecessor Benedict XVI likewise proposed "eliminating the structural causes of the dysfunctions of the world economy and correcting models of growth which have proved incapable of ensuring respect for the environment."[10] ...

7. These statements of the Popes echo the reflections of numerous scientists, philosophers, theologians and civic groups, all of which have enriched the Church's thinking on these questions. Outside the Catholic Church, other Churches and Christian communities—and other

---

    4. Encyclical Letter *Redemptor Hominis* (4 March 1979), 15: AAS 71 (1979), 287.
    5. Cf. *Catechesis* (17 January 2001), 4: *Insegnamenti* 41/1 (2001), 179.
    6. Encyclical Letter *Centesimus Annus* (1 May 1991), 38: AAS 83 (1991), 841.
    7. Ibid., 58: AAS 83 (1991), p. 863.
    8. John Paul II, Encyclical Letter *Sollicitudo Rei Socialis* (30 December 1987), 34: AAS 80 (1988), 559.
    9. Cf.id., Encyclical Letter *Centesimus Annus* (1 May 1991), 37: AAS 83 (1991), 840.
    10. Address to the Diplomatic Corps Accredited to the Holy See (8 January 2007): AAS 99 (2007), 73.

religions as well—have expressed deep concern and offered valuable reflections on issues which all of us find disturbing. . . .

10. I do not want to write this Encyclical without turning to that attractive and compelling figure, whose name I took as my guide and inspiration when I was elected Bishop of Rome. I believe that Saint Francis is the example par excellence of care for the vulnerable and of an integral ecology lived out joyfully and authentically. He is the patron saint of all who study and work in the area of ecology, and he is also much loved by non-Christians. He was particularly concerned for God's creation and for the poor and outcast. He loved, and was deeply loved for his joy, his generous self-giving, his openheartedness. He was a mystic and a pilgrim who lived in simplicity and in wonderful harmony with God, with others, with nature and with himself. He shows us just how inseparable the bond is between concern for nature and justice for the poor, commitment to society, and interior peace.

11. Francis helps us to see that an integral ecology calls for openness to categories which transcend the language of mathematics and biology, and take us to the heart of what it is to be human. . . .

12. What is more, Saint Francis, faithful to Scripture, invites us to see nature as a magnificent book in which God speaks to us and grants us a glimpse of his infinite beauty and goodness. "Through the greatness and the beauty of creatures one comes to know by analogy their maker" (*Wis* 13:5); indeed, "his eternal power and divinity have been made known through his works since the creation of the world" (*Rom* 1:20). . . .

13. The urgent challenge to protect our common home includes a concern to bring the whole human family together to seek a sustainable and integral development, for we know that things can change. . . . Young people demand change. They wonder how anyone can claim to be building a better future without thinking of the environmental crisis and the sufferings of the excluded.

14. I urgently appeal, then, for a new dialogue about how we are shaping the future of our planet. We need a conversation which includes every-

one, since the environmental challenge we are undergoing, and its human roots, concern and affect us all. . . .

As the bishops of Southern Africa have stated: "Everyone's talents and involvement are needed to redress the damage caused by human abuse of God's creation."[22] All of us can cooperate as instruments of God for the care of creation, each according to his or her own culture, experience, involvements and talents.

15. It is my hope that this Encyclical Letter, which is now added to the body of the Church's social teaching, can help us to acknowledge the appeal, immensity and urgency of the challenge we face. I will begin by briefly reviewing several aspects of the present ecological crisis, with the aim of drawing on the results of the best scientific research available today, letting them touch us deeply and provide a concrete foundation for the ethical and spiritual itinerary that follows. I will then consider some principles drawn from the Judeo-Christian tradition which can render our commitment to the environment more coherent. I will then attempt to get to the roots of the present situation, so as to consider not only its symptoms but also its deepest causes. This will help to provide an approach to ecology which respects our unique place as human beings in this world and our relationship to our surroundings. In light of this reflection, I will advance some broader proposals for dialogue and action which would involve each of us as individuals, and also affect international policy. Finally, convinced as I am that change is impossible without motivation and a process of education, I will offer some inspired guidelines for human development to be found in the treasure of Christian spiritual experience. . . .

23. The climate is a common good, belonging to all and meant for all. At the global level, it is a complex system linked to many of the essential conditions for human life. A very solid scientific consensus indicates that we are presently witnessing a disturbing warming of the climatic system. In recent decades this warming has been accompanied by a constant rise in the sea level and, it would appear, by an increase of extreme weather events, even if a scientifically determinable cause

---

22. Southern African Catholic Bishops' Conference, *Pastoral Statement on the Environmental Crisis* (5 September 1999).

cannot be assigned to each particular phenomenon. Humanity is called to recognize the need for changes of lifestyle, production and consumption, in order to combat this warming or at least the human causes which produce or aggravate it. It is true that there are other factors (such as volcanic activity, variations in the earth's orbit and axis, the solar cycle), yet a number of scientific studies indicate that most global warming in recent decades is due to the great concentration of greenhouse gases (carbon dioxide, methane, nitrogen oxides and others) released mainly as a result of human activity. As these gases build up in the atmosphere, they hamper the escape of heat produced by sunlight at the earth's surface. The problem is aggravated by a model of development based on the intensive use of fossil fuels, which is at the heart of the worldwide energy system. . . .

46. The social dimensions of global change include the effects of technological innovations on employment, social exclusion, an inequitable distribution and consumption of energy and other services, social breakdown, increased violence and a rise in new forms of social aggression, drug trafficking, growing drug use by young people, and the loss of identity. These are signs that the growth of the past two centuries has not always led to an integral development and an improvement in the quality of life. Some of these signs are also symptomatic of real social decline, the silent rupture of the bonds of integration and social cohesion. . . .

49. It needs to be said that, generally speaking, there is little in the way of clear awareness of problems which especially affect the excluded. Yet they are the majority of the planet's population, billions of people. These days, they are mentioned in international political and economic discussions, but one often has the impression that their problems are brought up as an afterthought, a question which gets added almost out of duty or in a tangential way, if not treated merely as collateral damage. . . .

At times this attitude exists side by side with a "green" rhetoric. Today, however, we have to realize that a true ecological approach *always* becomes a social approach; it must integrate questions of justice in debates on the environment, so as to hear *both the cry of the earth and the cry of the poor.*

50. Instead of resolving the problems of the poor and thinking of how the world can be different, some can only propose a reduction in the birth rate. At times, developing countries face forms of international pressure which make economic assistance contingent on certain policies of "reproductive health." Yet "while it is true that an unequal distribution of the population and of available resources creates obstacles to development and a sustainable use of the environment, it must nonetheless be recognized that demographic growth is fully compatible with an integral and shared development."[28] To blame population growth instead of extreme and selective consumerism on the part of some is one way of refusing to face the issues. . . .

61. On many concrete questions, the Church has no reason to offer a definitive opinion; she knows that honest debate must be encouraged among experts, while respecting divergent views. But we need only take a frank look at the facts to see that our common home is falling into serious disrepair. Hope would have us recognize that there is always a way out, that we can always redirect our steps, that we can always do something to solve our problems. Still, we can see signs that things are now reaching a breaking point, due to the rapid pace of change and degradation; these are evident in large-scale natural disasters as well as social and even financial crises, for the world's problems cannot be analyzed or explained in isolation. . . . "If we scan the regions of our planet, we immediately see that humanity has disappointed God's expectations."[35] . . .

66. The creation accounts in the book of Genesis contain, in their own symbolic and narrative language, profound teachings about human existence and its historical reality. They suggest that human life is

---

28. Pontifical Council for Justice and Peace, *Compendium of the Social Doctrine of the Church*, 483.

35. Id., *Catechesis* (17 January 2001), 3: *Insegnamenti* 24/1 (2001), 178.

grounded in three fundamental and closely intertwined relationships: with God, with our neighbour and with the earth itself. According to the Bible, these three vital relationships have been broken, both outwardly and within us. This rupture is sin. The harmony between the Creator, humanity and creation as a whole was disrupted by our presuming to take the place of God and refusing to acknowledge our creaturely limitations. This in turn distorted our mandate to "have dominion" over the earth (cf. Gen 1:28), to "till it and keep it" (Gen 2:15). As a result, the originally harmonious relationship between human beings and nature became conflictual (cf. Gen 3:17–19). It is significant that the harmony which Saint Francis of Assisi experienced with all creatures was seen as a healing of that rupture....

159. The notion of the common good also extends to future generations. The global economic crises have made painfully obvious the detrimental effects of disregarding our common destiny, which cannot exclude those who come after us. We can no longer speak of sustainable development apart from intergenerational solidarity. Once we start to think about the kind of world we are leaving to future generations, we look at things differently; we realize that the world is a gift which we have freely received and must share with others. Since the world has been given to us, we can no longer view reality in a purely utilitarian way, in which efficiency and productivity are entirely geared to our individual benefit. Intergenerational solidarity is not optional, but rather a basic question of justice, since the world we have received also belongs to those who will follow us. The Portuguese bishops have called upon us to acknowledge this obligation of justice: "The environment is part of a logic of receptivity. It is on loan to each generation, which must then hand it on to the next."[124] An integral ecology is marked by this broader vision....

246. At the conclusion of this lengthy reflection which has been both joyful and troubling, I propose that we offer two prayers. The first we can share with all who believe in a God who is the all-powerful Creator,

---

124. Portuguese Bishops' Conference, Pastoral Letter *Responsabilidade Solidária pelo Bem Comum* (15 September 2003), 20.

while in the other we Christians ask for inspiration to take up the commitment to creation set before us by the Gospel of Jesus.

A prayer for our earth

> All-powerful God,
> you are present in the whole universe
> and in the smallest of your creatures.
> You embrace with your tenderness all that exists.
> Pour out upon us the power of your love,
> that we may protect life and beauty.
> Fill us with peace, that we may live
> as brothers and sisters, harming no one.
> O God of the poor,
> help us to rescue the abandoned
> and forgotten of this earth,
> so precious in your eyes.
> Bring healing to our lives,
> that we may protect the world and not prey on it,
> that we may sow beauty,
> not pollution and destruction.
> Touch the hearts
> of those who look only for gain
> at the expense of the poor and the earth.
> Teach us to discover the worth of each thing,
> to be filled with awe and contemplation,
> to recognize that we are profoundly united
> with every creature
> as we journey towards your infinite light.
> We thank you for being with us each day.
> Encourage us, we pray, in our struggle
> for justice, love and peace.

A Christian prayer in union with creation

> Father, we praise you with all your creatures.
> They came forth from your all-powerful hand;

they are yours, filled with your presence and your tender love.
Praise be to you!

Son of God, Jesus,
through you all things were made.
You were formed in the womb of Mary our Mother,
you became part of this earth,
and you gazed upon this world with human eyes.
Today you are alive in every creature
in your risen glory.
Praise be to you!

Holy Spirit, by your light
you guide this world towards the Father's love
and accompany creation as it groans in travail.
You also dwell in our hearts
and you inspire us to do what is good.
Praise be to you!

Triune Lord,
wondrous community of infinite love,
teach us to contemplate you
in the beauty of the universe,
for all things speak of you.
Awaken our praise and thankfulness
for every being that you have made.
Give us the grace to feel profoundly joined
to everything that is.

God of love, show us our place in this world
as channels of your love
for all the creatures of this earth,
for not one of them is forgotten in your sight.
Enlighten those who possess power and money
that they may avoid the sin of indifference,
that they may love the common good,
advance the weak,

and care for this world in which we live.
The poor and the earth are crying out.
O Lord, seize us with your power and light,
help us to protect all life,
to prepare for a better future,
for the coming of your Kingdom
of justice, peace, love and beauty.
Praise be to you!
Amen.

    Given in Rome at Saint Peter's on 24 May, the Solemnity of Pentecost, in the year 2015, the third of my Pontificate.

INDEX

A

Abelson, Philip, 128
acid rain, 280
acidity of fresh waters, increased, 101
Ackerman, Tom P., 147, 195, 201
aerosols
    anthropogenic (tropospheric), 253
    in atmosphere, 140
    and atmospheric carbon dioxide, 161–163
aircraft exhaust, 112
Alito, Justice Samuel, 318
American Association for the Advancement of Science (AAAS), 86, 88, 128–131
American Clean Energy and Security Act of 2009, 269–270
animal grazing practices, 111
Antarctic ice
    cap, melting of, 100–101
    cores, analysis of air in, HD10–12
Anthropocene geological epoch, 282–286
Arrhenius, Svante, 9, 22, 23, 24, 39–44, 45, 46, 56
Assisi, Saint Francis of, 321, 324, 328
atmosphere, implications of rising carbon dioxide content of, 91–95
atmospheric carbon dioxide
    and aerosols, 161–163
    Arrhenius's model of, 39
    and climate, 175–179
    effects of, 99–101
    from fossil fuels, 99–100
    increase of, 55–59, 107
    increasing concentrations of, 183
    at Mauna Loa Observatory, HD2–4
    and methane and CFC-11 (fig.), HD14
    possible effects of, 100–102
    rate of accumulation of, 46–47
atmospheric research, 77–83

B

Bakerian Lecture, the, 32–38
banning coal, shale oil, 190, 191
Behind the Curve (Howe), 16
Behrens III, William W., 103
"Berlin Mandate," 265
Blair, Tony, 303
Blue Book of National Center for Atmospheric Research (NCAR), 77
Bolin, Bert, 50
Bradley, Raymond, HD9
Breakthrough Institute, 287
Broecker, Wallace, 96
Bronk, Detlev, 77
Brown, Gordon, 294
Brown, Harrison, 128
Brundtland, Gro Harlem, 220
Brundtland Report, the, 220–223, 224
Bryson, Reid, 147, 159, 164
Budyko, Dr. M.I., 156, 157
Burns, David, 88, 134–135

Bush, George H. W., 238, 239, 241, 242, 243, 256
Bush, George W., 287, 289, 294, 302, 307
Byrd, Senator Robert, 264
Byrd-Hagel Resolution on Kyoto Protocol, 213, 264–267

C

Callendar, G. S., 9, 24, 45, 49, 55, 56
Cannon, Jonathan Z., 312
carbon, social costs of, 298
carbon dioxide
   See also atmospheric carbon dioxide
   action flow, 132–133
   artificial production of, and influence on temperature, 45–47
   and climate, 175–179
   concentration profiles (figs.), HD16
   exchange between atmosphere and ocean, 55–59
   fuels and, 62
   and greenhouse effect, 68–69
   implications of rising content on atmosphere, 91–95
   increase of, 138, 181–190
   -induced global warming, 141–144
   research, 136–140
   warming caused by increased, 72–73
carbonic acid, 39–44
Care of our Common Home, On (Pope Francis), 321–331
Carson, Rachel, 96
Carter, President Jimmy, 86, 88, 134, 175, 311
Central Intelligence Agency (CIA)
   reports on climate change and food security, 147
   study of climatological research and intelligence, 151–160
Chakrabarty, Dipesh, HD6

Chamberlin, T. C., 56
Charney, Jule, 175, 181
Charney Report, 175, 180
chlorofluorocarbons (CFCs), 216, 240
Clean Air Act of 1990, 142, 238, 239, 245, 272, 273, 289, 308–320
Clean Development Mechanism, 301
Clean Water Act, 289
climate
   Advisory Group on Climate, 128–131
   carbon dioxide and, 175–179
   catastrophe, nuclear winter and, 197–200
   changing, 180–188
   forecasting, 67
   human influence on global, 254
   inadvertent modification of, 108–114
   man's activities influencing, 110–112
   National Climate Program Act of 1978, 123–127
climate change, 138, 180–188
   in Africa, 153
   CO_2-Induced Climatic Change: framework for policy choices (table), 185
   as controversy, 145–149
   economics of, 294–301, 302–306
   First Assessment Report (AR1) of Intergovernmental Panel on Climate Change, 215–219
   fossil fuels' effect on, 67–69
   governance, 209–214
   history, making, 3–18
   history of, 66–67
   human-induced, 295, 296–297
   from increased carbon dioxide, 101–102, 182–190
   the past, present, and future, 269–275
   a perspective on, 164–165

Stern Review on economics of, 211
United Nations Framework Convention on Climate Change (UNFCCC), 229–237
and uses of history, xi–xiv
Climate Change Treaty, 209, 243
climatology, current approaches to, 156–157
Clinton, Bill, 213, 264, 289
Club of Rome, 103–105, 108
CO_2-Induced Climatic Change: framework for policy choices (table), 185
coal, ban on, 191
Cold War
impact of weather control on, 70–76
and research expansion, 10, 11
roots of global warming, 49–52
colored pigments over polar ice surfaces, 72
Communism, 71
Community Climate (Systems) Model, 148
Connolley, William, 145
Conservative Foundation, the, 91, 92
Cosmos (tv show), 195
Craig, Harmon, 96
Crutzen, Paul J., 270, 272, 282, HD6

D

Darwin, Charles, 22
de Chardin, P. Teilhard, 283
"Death of Environmentalism, The" (Shellenberger & Nordhaus), 287–293
deforestation, 182
Discovery of Global Warming, The (Weart), 16
doomsday machine, 204

E

earth
"breathing," HD4, HD6
heat balance of, 64–65
temperature variations from years 1000 to 2100 (fig.), HD13
Earth in the Balance (Gore), 244
Earth Summit, 224, 229, 241, 243, 244
economics of climate change, 294–306
Ehrlich, P.R., 201
Eisenhower, President Dwight, 60, 67, 70–71
Eizenstat, Stuart, 256
El Niño effects, 143
emissions trading, 255–256
End of Nature, (McKibben), 14–15, 17, 270, 274
Endangered Species Act, 289
Energy and Climate (Revelle), 181
Energy Security Act of 1980, 181
environment
death of environmentalism, 287–293
"no regrets" environmental policy, 238–240
the past, present, and future, 269–275
restoring quality of our, 96–102
Rio Declaration on Environment and Development, 224–228
Sierra Club's five-year plan, 115–120
Environmental Protection Agency (EPA), 13, 141, 145, 149, 189, 194, 195, 307–320
environmentalism, death of, 287–293
Eriksson, Erik, 50

F

Fate of the Earth, The (Schell), 198
Federal Emergency Management Agency (FEMA), 202

Fleck, John, 145
food security, 151, 152–154
forecasting climate, 67
fossil fuels, 107, 284
    atmospheric carbon dioxide from, 99–100
    effect on climate change, 67–69
    residues in atmosphere, 94
Fourier, Jean-Baptiste Joseph, 9, 21–22, 23, 27, 40, 269
Friends of the Earth, 141

G

Galbraith, John Kenneth, 168
Gates, Dr. Larry, 159
General Circulation Model, 157, 159
Genesis Strategy, The (Schneider), 147, 166–173, 269
Global Climate Protection Act, 245
global commons, 120
global warming, 194, 270, 287–293, 308
    carbon dioxide-induced, 141–144
    cold war roots of, 49–52
    death of environmentalism, 287–293
    economics of, 294–302
    general remarks on, 27–31
    impacts of, 206–208
    making global warming green, 85–88
    Roger Revelle and, 52
    scientific "prehistory" of, 21–24
Gore, Al, 12, 136–140, 141, 213, 241, 242–244, 245–246, 256, HD9
governance of climate change, 209–214
grain crisis, 155, 158
Gray, C. Boyden, 238, 239
greenhouse
    how to live in a, 193–194
    effect, 28, 208, 216, 232
    gases, 184, 190, 217, 218, 230, 240, 252–253, 257, 262, 265, 284, 298, 301, 307
    warming, 68–69, 185, 189–193, 194

H

Hagel, Senator Chuck, 264
Hansen, James, 206, 215, 277
heat balance of earth, 64–65
historicizing data, HD1–19
history
    climate change and uses of, xi–xiv
    making climate change, 3–18
    scientific "prehistory" of global warming, 21–24
Holocene ("Recent Whole"), 283
Hornig, Donald, 97
hot summers, 208
Houghton, Henry G., 78
Howe, Joshua, xii, xiii, xiv
Hughes, Malcolm, HD9
human ecology, 323
hydrochlorofluorocarbons, 240

I

Ice Age, 42–44, 73
ice ages, 162, 163, 165
intelligence problems, study of climatological research and, 151–160
intelligence problems, study of climatological research as it pertains to, 155
Intergovernmental Panel on Climate Change (IPCC), 14, 209–210, 213, 220, 248
    CO_2 stabilization scenarios (fig.), HD19
    described, 8
    First Assessment Report (AR1), 215–219
    Second Assessment Report (SAR), 247–254
    socio-economic scenarios (fig.), HD17–18
    Third Assessment Report (TAR), HD15

international campaigns, criteria for, 121–122
International Geophysical Year (IGY), 51, 70, 77, 83
International Monetary Fund, 220
Irish great potato famine, 158
Isaacs, Dr. John, 159

J

Johnson, Lyndon, 52, 73–74, 85
Joseph, Biblical story of, 166–167, 170

K

Kahn, Herman, 204
Keeling, Charles David, 11, 85, 91, 96, HD5, HD7
Keeling Curve, 16, 91, HD1–19
Kennedy, Senator Edward, 202
Kerry, John, 242
Keyes, Dale, 188
Khrushchev, Nikita, 153
Kratatoa volcano, 73
Kutzbach, John, 159
Kyoto Protocol, 211, 212–213, 264, 266, 302
   Byrd-Hagel Resolution on, 264–267
   to United Nations Framework Convention on Climate Change (UNFCCC), 255–263

L

lakes, artificial, 111
Lamb, Professor H.H., 156
lamp black, 72
Landsberg, Helmut E., 147, 166, 169, 172–174
Larson v. Valente, 314–315
Laudato Si' (*On Care for Our Common Home* (Pope Francis), 15, 271, 319, 321–331

Le Roy, E., 283
Limits of Growth, The (the Club of Rome), 103–107
limits to growth, 103–105
Little Ice Age, 152, 158
Lovelock, James, 280
Lujan v. Defenders of Wildlife, 319
Lyell, Sir Charles, 283

M

Making Climate Change History (Howe), XII, XIII, 3–18
Malik, Dr. Charles H., 82
Manabe, Suki, 106, 207
Mann, Michael, HD9, HD12
Margulis, Lynn, 280
Markin, Arkady, Borisovich, 74
Mars, 65, 162
Marsh, G. P., 283
Massachusetts v. Environmental Protection Agency, 15, 271–272, 273, 307–320
Mauna Loa Observatory, atmospheric carbon dioxide at, HD2–5
McCloskey, Michael, 86, 88, 115, 121
McConnell, Mitch, 241, 244–246
McKibben, Bill, 14–15, 270, 271, 274, 277, 282
Meadows, Dennis L., 103, 105
Meadows, Donella H., 103
Mesirow, Lynn E., 172
meteorology, 156
models, climate, 113
Montreal Protocol on Substances that Deplete the Ozone Layer (1987), 231, 234, 257, 260
Moss, Tom, 128
Mt. Pinatubo volcanic eruption, 1991, 253
multilateralism, 222

N

Namais, Dr. Jerome, 159
NASA Goddard Institute for Space Studies, 207
National Academy of Sciences, 13, 141, 145, 149, 175, 176–178, 180, 189, 194, 195
National Center for Atmospheric Research (NCAR), 77–83, 148
National Climate Policy Act, 136
National Climate Program Act of 1978, 134, 311
National Energy Strategy (NES) legislation, 239
National Environmental Policy Act, 289
National Institute for Atmospheric Research, 79–83
National Science Foundation, 77, 78, 81–82
Nature, The End of (McKibben), 277–281
New York Times, 193–194
Nierenberg, William, 180, 189
Nierenberg Report for the National Academy of Sciences, 189, 294–295
Nixon, Richard, 76
Nordhaus, Ted, 15, 270, 273, 287
Nordhaus, William D., 180, 194, 294–295, 302
nuclear technology, 167
nuclear winter
 about, 201–202
 and climatic catastrophe, 197–200
 global consequences of multiple nuclear explosions, 195–196
 reappraised, 203–205
 term's meaning, 147

O

Oberth, Dr. Hermann, 73
Oblast, Lvov, 155
oceans
 carbon dioxide exchange between atmosphere and, 55–59
 situation of, 60–63
Office of Science and Technology Policy, 180
On Care for Our Common Home (Pope Francis), 15, 271, 319, 321–331
On the Beach (Shute), 198
Orville, Howard T., 52, 70, 77
ozone layer, 222, 231, 284–285

P

Peccei, Dr. Aurelio, 104
Peterson, Thomas, 145
Pewitt, N. Douglas, 136–140
photosynthesis, increase with increasing carbon dioxide, 101
policies, energy and nonenergy, 191
Pollack, J. B., 147, 195, 201
pollution, 96–99, 107, 307
Pomerance, Rafe, 141–144
Pope Benedict XVI, 323
Pope Francis, 15, 271, 321–331
Pope Paul II, 323
Pope Paul VI, 322
Pope Saint John XXIII, 322
power plants, scrubbing CO_2 emissions from, 192, 194
prayers, 329–331
precipitation, modifying by seeding clouds, 111

R

radioactive fallout, 203
Randers, Jørgen, 103
Rasool, S. Ishtiaque, 161, 162, 166
ratio photospectrometer (fig.), 38
Reagan, Ronald, 88, 134, 136
Reilly, William, 240, 241

Restoring the Quality of Our Environment (Science Advisory Council), 85
Revelle, Roger, 10, 11, 49, 50, 51, 55, 60–63, 64–69, 77, 85, 91, 96, 128, 140, 180, HD5
Rio Declaration on Environment and Development, 224–228, 236, 241
Rio Earth Summit, 241–246
rivers, diversion of, 111
Rivkin Jr., David B., 238, 239
Robert, Chief Justice John, 271, 272, 307, 318
Russians, 62–63
Russia(ns), 74, 75, 158

S

Sagan, Carl, 147–148, 149, 195–196, 197, 201, 203
Santer, Ben, 247
satellites
    See also Sputnik satellite
    as platforms for weather control, 73–74
Saussure, M. de, 29
Scalia, Justice Antonin, 307, 317n, 318
Scharlin, Patricia, 87, 115
Schell, Jonathan, 198
Schellenberger, Michael, 15, 270, 273
Schelling, Thomas C., 180, 181, 186, 194
Schneider, Stephen H., 147, 148, 161, 162, 166, 170, 171, 172–174, 203, 269
scrubbing CO_2 emissions from power plants, 192, 194
sea level rise, 100–101, 218
SEC v. Chenery Corp., 314
Seidel, Stephen, 189
Sellers, Dr. William, 159
Shabecoff, Philip, 207
shale oil, ban on, 191
Shellenberger, Michael, 287
Shute, Neil, 198
Sierra Club, the, 11–12, 86, 87, 115

Silent Spring (Carson), 96
Singer, S. Fred, 201, 202, 203
Slade, David, 86, 87, 88, 132, 134–135
Smagorinsky, Joseph, 96, 151, 156, 174
smog, 112
Sputnik satellite, 52, 71, 76
Stern, Sir Nicholas, 15, 271, 294, 303
Stern Review on Economics of Climate Change, 211, 294–306
Stevens, Justice John Paul, 308
Stewart, Balfour, 37
Stoermer, Eugene F., 270, 272, 282, HD6
Stoppani, Antonio, 283
Study of Critical Environmental Problems (SCEP), 109
Study of Man's Impact on Climate (SMIC), 108–114
Suess, Hans E., 10, 49, 50, 52, 55, 60, 85
summaries for policymakers (SPMs) of IPCC, 210, 249, 250
Suomi, Verner E., 177
Supreme Court, 307
sustainable development, 212, 223, 238
Sutter, Paul S., xi
Synthetic Fuels Corporation, 143

T

Tao Ti Ching, 290
Teller, Dr. Edward, 72, 73, 75
temperature
    See also warming
    artificial production of carbon dioxide and influence on, 45–47
    global mean surface air, 218, 253
    of the globe, 27–31
    of the ground, influence of carbonic acid in air upon, 39–44
    water, 100–101
thermometers, 30
Thomas, Albert Richard, 52, 60–62, 64–69

Thomas, Justice Clarence, 318
Thompson, Starley L., 203
Thompson, William, 22
Toon, Brian, 147
Toon, Brian O., 201
Toon, O. B., 195
transportation and climatic modification, 111–112
Truman, President Harry, 306
TTAPS paper summary, 195–196
Turco, R. P., 195
Turco, Richard P., 147, 201
Tyndall, John, 9, 21–22, 23, 24, 32, 39, 40, 45, 269

U

United Nations, 224–228, 229, 255
United Nations Charter, 230
United Nations Conference on Environment and Development (UNCED) in Rio de Janeiro, Brazil, 212, 224, 229
United Nations Framework Convention on Climate Change (UNFCCC), 209, 211–213, 229–237, 238, 251, 255
    Kyoto Protocol to, 255–263
    Second Assessment Report (SAR), 247–254
United Nations' World Commission on Environment and Development (WCED), 212
University Committee on Atmospheric Research (UCAR), 78
University of Wisconsin study on climatic change, 157–160
urbanization, 284
U.S. Carbon Dioxide Research and Assessment Program, 132–133
U.S. Central Intelligence Agency. *See* Central Intelligence Agency (CIA)
U.S. Department of Defense, 205
U.S. Weather Bureau, 310

V

Venus, 162
Vernadsky, V. I., 283
Vienna Convention for the Protection of the Ozone Layer, 1985, 231, 243
volcanoes, 57, 58–59, 73, 196

W

warming
    caused by increased carbon dioxide, 72–73
    of the earth, 62, 177
    global. *See* global warming
    greenhouse, 189–193
    of sea water, 100–101
Waterman, Dr. Alan T., 78
Watts, William, 104
Weart, Spencer, 16
weather control, impact on cold war, 70–76
Wexler, Harry, HD5
White, Robert, 128
wildness, 279–280
Williamson v. Lee Optical of Okla., Inc., 314
Wirth, Senator Tim, 106
Wittwer, Sylvan, 128
World Bank, 220
World Commission on Environment and Development, 14, 220

Y

Yates, Sidney, 52, 60

WEYERHAEUSER ENVIRONMENTAL BOOKS

Defending Giants: The Redwood Wars and the Transformation of American Environmental Politics, by Darren Frederick Speece

The City Is More Than Human: An Animal History of Seattle, by Frederick L. Brown

Wilderburbs: Communities on Nature's Edge, by Lincoln Bramwell

How to Read the American West: A Field Guide, by William Wyckoff

Behind the Curve: Science and Politics of Global Warming, by Joshua P. Howe

Whales and Nations: Environmental Diplomacy on the High Seas, by Kurkpatrick Dorsey

Loving Nature, Fearing the State: Environmentalism and Antigovernment Politics before Reagan, by Brian Allen Drake

Pests in the City: Flies, Bedbugs, Cockroaches, and Rats, by Dawn Day Biehler

Tangled Roots: The Appalachian Trail and American Environmental Politics, by Sarah Mittlefehldt

Vacationland: Tourism and Environment in the Colorado High Country, by William Philpott

Car Country: An Environmental History, by Christopher W. Wells

Nature Next Door: Cities and Trees in the American Northeast, by Ellen Stroud

Pumpkin: The Curious History of an American Icon, by Cindy Ott

The Promise of Wilderness: American Environmental Politics since 1964, by James Morton Turner

The Republic of Nature: An Environmental History of the United States, by Mark Fiege

A Storied Wilderness: Rewilding the Apostle Islands, by James W. Feldman

Iceland Imagined: Nature, Culture, and Storytelling in the North Atlantic, by Karen Oslund

Quagmire: Nation-Building and Nature in the Mekong Delta, by David Biggs

Seeking Refuge: Birds and Landscapes of the Pacific Flyway, by Robert M. Wilson

Toxic Archipelago: A History of Industrial Disease in Japan, by Brett L. Walker

Dreaming of Sheep in Navajo Country, by Marsha L. Weisiger

Shaping the Shoreline: Fisheries and Tourism on the Monterey Coast, by Connie Y. Chiang

The Fishermen's Frontier: People and Salmon in Southeast Alaska, by David F. Arnold

Making Mountains: New York City and the Catskills, by David Stradling

Plowed Under: Agriculture and Environment in the Palouse, by Andrew P. Duffin

The Country in the City: The Greening of the San Francisco Bay Area, by Richard A. Walker

Native Seattle: Histories from the Crossing-Over Place, by Coll Thrush

Drawing Lines in the Forest: Creating Wilderness Areas in the Pacific Northwest, by Kevin R. Marsh

Public Power, Private Dams: The Hells Canyon High Dam Controversy, by Karl Boyd Brooks

Windshield Wilderness: Cars, Roads, and Nature in Washington's National Parks, by David Louter

On the Road Again: Montana's Changing Landscape, by William Wyckoff

Wilderness Forever: Howard Zahniser and the Path to the Wilderness Act, by Mark Harvey

The Lost Wolves of Japan, by Brett L. Walker

Landscapes of Conflict: The Oregon Story, 1940–2000, by William G. Robbins

Faith in Nature: Environmentalism as Religious Quest, by Thomas R. Dunlap

The Nature of Gold: An Environmental History of the Klondike Gold Rush, by Kathryn Morse

Where Land and Water Meet: A Western Landscape Transformed, by Nancy Langston

The Rhine: An Eco-Biography, 1815–2000, by Mark Cioc

Driven Wild: How the Fight against Automobiles Launched the Modern Wilderness Movement, by Paul S. Sutter

George Perkins Marsh: Prophet of Conservation, by David Lowenthal

Making Salmon: An Environmental History of the Northwest Fisheries Crisis, by Joseph E. Taylor III

Irrigated Eden: The Making of an Agricultural Landscape in the American West, by Mark Fiege

The Dawn of Conservation Diplomacy: U.S.-Canadian Wildlife Protection Treaties in the Progressive Era, by Kirkpatrick Dorsey

Landscapes of Promise: The Oregon Story, 1800–1940, by William G. Robbins

Forest Dreams, Forest Nightmares: The Paradox of Old Growth in the Inland West, by Nancy Langston

The Natural History of Puget Sound Country, by Arthur R. Kruckeberg

WEYERHAEUSER ENVIRONMENTAL CLASSICS

Making Climate Change History: Primary Sources from Global Warming's Past, edited by Joshua P. Howe

Nuclear Reactions: Documenting American Encounters with Nuclear Energy, edited by James W. Feldman

The Wilderness Writings of Howard Zahniser, edited by Mark Harvey

The Environmental Moment: 1968–1972, edited by David Stradling

Reel Nature: America's Romance with Wildlife on Film, by Gregg Mitman

DDT, Silent Spring, and the Rise of Environmentalism, edited by Thomas R. Dunlap

Conservation in the Progressive Era: Classic Texts, edited by David Stradling

Man and Nature: Or, Physical Geography as Modified by Human Action, by George Perkins Marsh

A Symbol of Wilderness: Echo Park and the American Conservation Movement, by Mark W. T. Harvey

Tutira: The Story of a New Zealand Sheep Station, by Herbert Guthrie-Smith

Mountain Gloom and Mountain Glory: The Development of the Aesthetics of the Infinite, by Marjorie Hope Nicolson

The Great Columbia Plain: A Historical Geography, 1805-1910, by Donald W. Meinig

CYCLE OF FIRE

Fire: A Brief History, by Stephen J. Pyne

The Ice: A Journey to Antarctica, by Stephen J. Pyne

Burning Bush: A Fire History of Australia, by Stephen J. Pyne

Fire in America: A Cultural History of Wildland and Rural Fire, by Stephen J. Pyne

Vestal Fire: An Environmental History, Told through Fire, of Europe and Europe's Encounter with the World, by Stephen J. Pyne

World Fire: The Culture of Fire on Earth, by Stephen J. Pyne

Also available:

Awful Splendour: A Fire History of Canada, by Stephen J. Pyne